Wissenschaftliche Beiträge
zur Medizinelektronik

Band 5

Wissenschaftliche Beiträge zur Medizinelektronik

Band 5

Herausgegeben von
Prof. Dr. Wolfgang Krautschneider

Implant System for the Recording of Internal Muscle Activity to Control a Hand Prosthesis

Vom Promotionsausschuss der

Technischen Universität Hamburg-Harburg

zur Erlangung des akademischen Grades

Doktor-Ingenieur (Dr.-Ing.)

genehmigte Dissertation

von

Lait Abu Saleh

aus

Majdal Shams, Golanhöhen

2015

1. Gutachter: Prof. Dr.-Ing. Wolfgang H. Krautschneider

2. Gutachter: Prof. Dr.-Ing. Jörg Müller

Tag der mündlichen Prüfung: 07. Oktober 2015

Lait Abu Saleh

Implant System for the Recording of Internal Muscle Activity to Control a Hand Prosthesis

Logos Verlag Berlin

λογος

Wissenschaftliche Beiträge zur Medizinelektronik

Herausgegeben von
Prof. Dr. Wolfgang Krautschneider

Technische Universität Hamburg-Harburg
Institut für Nano- und Medizinelektronik
Eißendorfer Str. 38
D-21073 Hamburg

Bibliografische Information Der Deutschen Bibliothek

Die Deutsche Bibliothek verzeichnet diese Publikation in der Deutschen
Nationalbibliografie; detaillierte bibliografische Daten sind im Internet
über http://dnb.ddb.de abrufbar.

ISBN 978-3-8325-4153-8
ISSN 2190-3905

Logos Verlag Berlin GmbH
Comeniushof, Gubener Str. 47,
10243 Berlin
Tel.:   +49 (0)30 / 42 85 10 90
Fax: +49 (0)30 / 42 85 10 92
http://www.logos-verlag.de

To

Lina, Dema and Hauke

Danksagung

Zu großem Dank bin ich Prof. Dr.-Ing. Wolfgang Krautschneider verpflichtet. Ohne seinen wertvollen akademischen Rat und seine Unterstützung wäre diese Arbeit nicht entstanden, zudem hat er entscheidenden Anteil daran, dass mir das Bleiberecht in Deutschland zuerkannt wurde.

Ein besonderer Dank gilt auch Dr.-Ing. habil. Dietmar Schröder. Immer wieder hat er mich mit Hinweisen aus seinem unerschöpflichen Fundus an thematischer und wissenschaftlicher Erfahrung in neue Sphären gelenkt.

Ebenso geht mein Dank an meine ehemaligen Kommilitonen und Mitarbeiter des Instituts für Nano- und Medizinelektronik für die Unterstützung im Rahmen des Projektes "MyoPlant". Sie haben mir in den vergangenen Jahren durch ihre Diskussionsbeiträge und wertvollen Tipps das wissenschaftliche Themenfeld erweitert und meinen Blick geschärft.

Herrn Prof. Dr.-Ing. Jörg Müller danke ich für die Zweitbegutachtung sowie Herrn Prof. Dr.-Ing. Arne Jacob für die Übernahme des Prüfungsvorsitzes.

Herzlich danke ich Pastor Hauke Schröder für seinen väterlichen Beistand seit meiner Ankunft in Deutschland und seine Freundschaft.

Die sprachliche Korrektur dieser Arbeit verdanke ich Frau Christiane Schröder.

Schließlich gedenke ich meiner Eltern, Schwiegereltern und Geschwister auf den besetzten und umkämpften Golanhöhen. Ich danke ihnen für ihre Liebe und ihr Vertrauen.

Besondere Unterstützung erfuhr ich durch meinen Bruder Dr. Aysar Abo Saleh.

Mein größter und tief empfundener Dank gilt meiner lieben Frau Lina.

Lait Abu Saleh, November 2015

Abstract

An implantable system for controlling a bionic hand prosthesis is presented. For this study an implantable system is designed to invasively acquire muscle activity. The specification of the implantable system is the result of the collaboration of physicians and engineers in a research project to develop a bionic hand prosthetic system by using an electronic implant. Multi-channel EMG is acquired by the implant system through epimysial electrodes. The system utilizes two wireless interfaces for data and power transmission. Furthermore, a custom made low-power application specific integrated circuit (ASIC) is designed in 130 nm technology to amplify, filter and digitize the analogue muscle-activity. A trade-off between power consumption, area and noise was considered during the design phase. The implant system was successfully tested in sheep and in rhesus macaques. Valuable EMG data was gathered by the implant system. Several mathematical methods were developed to evaluate the quality of the internally recorded EMG. A surface EMG recording was initiated parallel to the internal one for comparison purposes. The calculations show that despite the low number of bits and the low data rate of the implant recording system, the SNR and the information rate of the muscle activity-regions were higher compared to the surface recorded EMG. Furthermore, the internal recorded amplitudes are twice as big as the external ones. The selectivity of the internal recording seems to be higher than its external counterpart due to the short duration of muscle activity regions compared to the long compound of muscle activity recorded on the body surface. This provides an opportunity for a simple and smooth control of a hand prosthetic system with high degrees of freedom.

Bionic Hand Prosthesis, Implant, Epimysial Electrodes, ASIC, Wireless Power Transmission, Wireless Data Transmission, Invasive EMG-Recording.

Contents

Glossary

Symbols

A_c — Common gain of an amplifier

A_d — Differential gain of an amplifier

A_o — The open-loop gain of an amplifier

A_{CL} — Closed loop gain

A_{OL} — Open loop gain

CMRR — Common mode rejection ratio

C_W — Warburg capacitance

C_{DL} — Double layer capacitance

C'_{ox} — Gate-oxide capacitance per unit area

DNL — Differential nonlinearity

DR — $DR = 10^{\frac{6.02*N+1.76}{10}}$ For an N bit ADC, this is the maximum signal energy versus the minimum step size energy

DR — Dynamic range

ENOB — Effective number of bits

ER — Effective resolution (bit)

F — Faraday constant ($96\,485.336\frac{C}{mol}$)

GBW — Gain bandwidth

INL — Integral nonlinearity

I_b — Bias current

I_d — Drain current

I_n^2 — Total current noise power

PM — Phase margin

PSD — Power spectral density

P — Power

Q_e — Quantization error

R_f — Faradic resistance

R_{DC} — Correction resistance

R_{HF} — Electrolyte resistance

R_{on} — The MOSFET on resistance

R — Gas constant ($8.314\frac{J}{molK}$)

SINAD — Signal-to-noise and distortion ratio

SNR_{dB} — $SNR_{dB} = 10\,Log_{10}(SNR)$

SNR — $SNR = \frac{S_{ignalrms}^2}{V_{noise}^2}$ Signal-to-noise ratio

THD — Total harmonic distortion

T — Temperature in Kelvin

V_m — The membrane potential of a nerve cell, internal potential minus external potential

V_n^2 — Total voltage noise power

V_{GS} — Gate-source voltage

V_{dd} — Supply voltage

V_{hc} — Half-cell potential

V_{os} — The input referred offset voltage

V_{pp} — Peak-to-peak voltage

V_{ref} — The reference voltage of an ADC

V_{rms} — The root mean square voltage: $V_{rms} = \frac{1}{\sqrt{2}}\left(\frac{V_{pp}}{2}\right)$

V_{th} — Threshold voltage

Z_{ETI}	Impedance of the electrode-tissue interface	**EMF**	Electromagnetic fields
		EMG	Electromyography
$\Sigma\Delta$ ADC	Sigma-Delta analog to digital converter	**EMI**	Electromagnetic interference
		EP	Evoked potentials
σ^2	Variance	**ETI**	Electrode-tissue interface
σ	Standard deviation	**FS**	Full scale is the maximum or minimum representative range of the ADC
c	Concentration (mol/L)		
g_d	$g_d = \frac{\partial I_d}{\partial V_{ds}}$ MOSFET output conductance		
		IC	Integrated circuit
g_m	$g_m = \frac{\partial I_d}{\partial V_{gs}}$ MOSFET transconductance	**IEMG**	Internal electromyography
		IMES	Implantable myoElectric sensor
i_n	RMS noise current	**LSB**	Least significant bit
k	Boltzmann constant $(1.38 \cdot 10^{-23} J/K)$	**MIM**	Metal-insulator-metal capacitance
		MOM	Metal-oxide-metal capacitance
q	Electron charge $(1.602 \cdot 10^{-19} C)$	**MOSFET**	Metal-oxide semiconductor field effect transistor
sps	samples per second		
v_n	RMS noise voltage	**MSB**	Most significant bit

Abbreviations

		MU	Motor unit: all muscle cells innervated by one nerve cell
ACF	Autocorrelation function		
ACh	Acetylcholine: A neurotransmitter used in body for excitation or inhibition purposes	**MUAP**	Motor unit action potential
		OTA	Operational transconductance amplifier
ADC	Analog-to-digital converter		
ASIC	Application specific integrated circuit	**PCB**	Printed circuit board
		RMS	Root mean square
Ag/AgCl	Silver/Silver-chloride	**RoA**	Regions of activity
CI	Coherent interference	**RoN**	Regions of noise
CMOS	Complementary metal-oxide semiconductor	**SAR**	Successive approximation register
		SEMG	Surface electromyography
DAC	Digital-to-analog converter	**SFDR**	Spurious-free dynamic range
DoF	Degrees of freedom	**SFEMG**	Single-fiber electromyography
ECG	Electrocardiography	**TG**	Transmission gate
EEG	Electroencyphalography	**TMR**	Trans-muscular re-innervation

List of Figures

List of Tables

Chapter 1

Introduction

Losing an upper extremity is a crucial incidence causing a huge restriction for amputees in daily life. There are 64 332 amputees in Germany, 15 273 lost one arm and 2137 lost both arms according to the statistics of the Bundesamt from 2011 [7]. In the USA there are 1.9 million amputees and approximately 185 000 amputation surgeries performed each year, 82 % of which are performed due to Peripheral Vascular Disease and Diabetes [8]. The loss of an upper extremity is also due to accidents, war or to amputation caused by trauma, disease, tumor or it maybe congenital [9]. Casualties are provided in most of the cases with a state of the art artificial hand prosthetic. The measurement of muscle activity remains the source for controlling most of the state of the art hand prosthetic systems nowadays [10]. Surface electromyogram (sEMG) is often used by placing surface electrodes on the skin overlying the target muscle. The control mechanism of a hand prosthesis employs independent source signals for achieving certain degrees of freedom (DoF) in the motion of the prosthetic hand. Generally, two independent signals (or rather independent signal components) are needed to accomplish one DoF. For example, to perform a hand grip (open or close the fingers for holding a certain object with one's hand) using a hand prosthesis, two independent control signals are needed: the first signal is for closing and the second is for opening the hand. Those signals can be acquired, for example, from the Flexor and Extensor muscles of the remaining part of the arm.

A surface EMG signal contains generally accumulated activity of many motor units from the underlying recording region and less from deep muscle regions. Besides, the high frequency components of muscle activity is smoothed in surface EMG-recordings due to the low-pass filter-characteristic of the skin layers. Therefore, most of the spectral power of the sEMG is located below 400-500 Hz which leads to a sampling rate of approximately 1 kHz [11], [12]. Cross-talk and interference from neighboring muscles greatly add to the surface-recorded signal according to the relatively big recording area covered by the surface electrode. Interference from several muscle groups also reduces the selectivity of sEMG.

Long-term sEMG recording leads to skin irritations caused by surface electrodes. The comfort of the hand prosthesis is hence affected. Furthermore, artifacts are a major concern in sEMG recording. Movement artifacts initiated by electrode movement relative to the skin or cable movement [13], electromagnetic coupling (50/60 Hz), and variable impedance of the skin interface caused by sweat are some of the issues affecting long-term signal quality and reliability of an external EMG.

Purpose of the work

This work describes a centralized implantable approach for acquiring internal muscle activity to overcome the disadvantages of an external recording as explained before and hence improve recorded EMG signal quality and reliability. Invasive EMG is less vulnerable to electrode movement artifacts and offers an opportunity for recording from deeper muscle regions [14]. Furthermore, measuring independent control-signals is essential for a prosthetic hand with multiple degrees of freedom (DoF). Therefore, an invasive recording would possibly offer a high selective EMG for advanced and smooth control of a hand prosthesis. Besides surface EMG possesses higher path impedance than invasive EMG due to the presence of more tissue layers and the skin layers as depicted in Fig. 1.1. This results in higher internally recorded amplitudes than the externally recorded ones. A parallel internal/external recording using our custom made implant system and a commercial surface recording system confirms the amplitude difference as seen in Fig. 1.2.

The implantable system aims at improving the signal stability and robustness and supply the prosthetic control unit with the needed EMG-activity over time in long-term recordings. It consists of an external part for wireless transmission of energy and data and an implantable electronic part. The implantable electronic system is capable of recording internal muscle activity using bipolar epimysial electrodes. It is divided into several parts or modules. The first module is the electrode

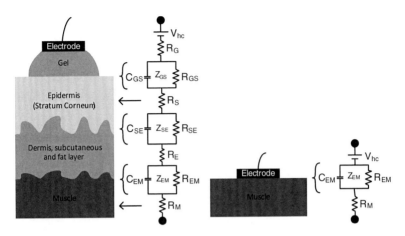

Figure 1.1: External and internal path impedance between muscle activity and the recording electrode.

interface followed by an electronic implant system for biopotential signal conditioning. An external module is used for wireless power and data transmission. The electrode interface utilizes epimysial silicon electrodes to acquire localized muscle activity. The electronic system contains an application specific integrated circuit (ASIC) for signal conditioning (amplifying, filtering, and digitizing biopotential signals), an energy-management circuit, a transceiver for wireless data transmission to an external receiver, a digital controller to coordinate data transmission on the electronic system and to control the communication with an external application. An external energy coil is used to power the implant electronic system per inductive coupling under dynamic movement conditions where the distance and angle between external and internal coils varies due to body movement (this is only required for animal experiments).

The main requirements for implantable electronics are compiled and discussed. A design concept is introduced according to a well defined specification based on the expertise of engineers, surgeons, electrode and prosthetic developers. The design of an ASIC for implantable application is described. It is based on a trade-off between power consumption, area, and noise. A system architecture for the implant electronic is proposed and discussed.

This work is the result of the Bundesministerium für Bildung und Forschung

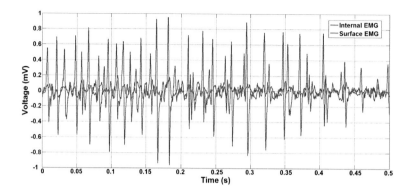

Figure 1.2: Internal and external EMG recorded from the biceps-muscle (amplitude comparison).

(BMBF) research project to develop a bionic hand prosthetic system using an intelligent implant (project number: 16SV3699).

Outline

This work is structured as follows: Chapter 2 describes some biomedical fundamentals and encloses some basic knowledge of system components for implantable biomedical application. A brief description of the development of upper extremity prosthesis is added to the end of this chapter. The architecture of the implant system is explained in Chapter 3. The development of the main component in the implant electronic system (MyoC1: the ASIC) is illustrated in Chapter 4. Chapters 5 and 7 describe the wireless telemetry and the data transmission protocol between the implant and the external interface. The implant electronic system is explained in Chapter 6. Chapter 8 illustrates the experimental results from animal models and some algorithms for the analysis of the internal recorded muscle activity compared to surface EMG. Finally, a work summary and a future outlook are given in Chapter 9.

Chapter 2

Fundamentals

Bio-signal acquisition is of vital importance for medical applications. Most diagnostic routines make use of bio-signal recordings for biological process monitoring and cure purposes. Furthermore, bio-signals are frequently used in controlling prosthetic systems to partly compensate the loss of body extremities. This chapter presents a brief look into the origin of biological signals and the instrumentation used for recording these signals. Moreover, it presents the most common bio-signal recordings in medical applications.

2.1   Origin of Bio-Signals

The origin of bio-signals lies in the electrical nature of signal transmission in nerve and muscle cells (Fig. 2.1). Bio-signals are electrical signals emerging from nerve and muscle cells. These cells possess a concentration gradient of certain ion species (Sodium "Na+", potassium "K+", calcium "Ca++", chloride "Cl-", see Fig. 2.2) between the cytoplasm (inner of the cell) and the extracellular space (outside of the cell). The cell membrane concludes a huge number of ion channels making it permeable for certain ion species. The ion channels are controlled by the voltage across the cell membrane. They show a time and voltage dependent characteristic. Diffusion (due to ion concentration gradient) and electrical force (due to the charge of ion species) result in voltage difference between the cytoplasm and the extra cellular space (membrane voltage V_m). The membrane voltage of the nerve cell is best described by the Goldman-Hodgkin-Katz Equation:

$$V_m = -\frac{RT}{F} \ln \frac{P_K c_{i,K} + P_{Na} c_{i,Na} + P_{cl} c_{o,Cl}}{P_K c_{o,K} + P_{Na} c_{o,Na} + P_{cl} c_{i,Cl}} \qquad (2.1)$$

where R is the gas constant (Joules per Kelvin per mole), T is the temperature (in Kelvin), F is the Faraday's constant (Coulombs per mole), P_{ION} is the permeability of the ion species (in meters per second), $c_{i/o,ION}$ is the intracellular/extracellular ion concentration (moles per cubic meter), and V_m depicts the potential difference between the cytoplasm and the extracellular volume (in Volts). In order to measure the membrane potential a laboratory setup is usually required. Using a thin needle electrode inserted into the cytoplasm and another thin electrode at the extracellular medium allow to measure the potential difference over the membrane of the cell and furthermore to control it in order to figure out the currents flowing through the

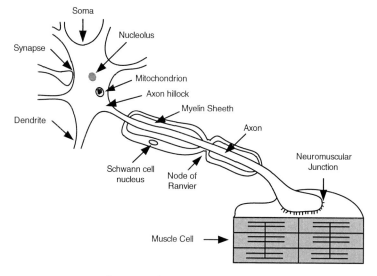

Figure 2.1: Structure of Nerve Cell [1].

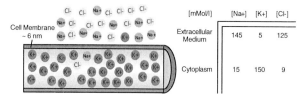

Figure 2.2: Ion concentrations in a typical mammalian nerve and in the extracellular medium of the nerve [2].

membrane (voltage clamp experiment). Thin glass micro-electrodes (micropipette, with a diameter of 1 μm) enable the measurement of certain ion channel currents through the membrane of the biological cell. Ionic currents in the range of several pico Ampere can be measured. In case, the membrane voltage reaches a certain threshold value (as a result of stimulation) it triggers an impulse called "action potential" (Fig. 2.3) through the cell. Nerve cells remit action potentials through synapses to the next nerve cell whereas the action potentials travel from a nerve cell to a muscle cell through a neuro-muscular junction. Neurotransmitter such as Acetylcholine (ACh) play a central role in forwarding action potentials between nerve cells as well as between nerve and muscle cells. According to the type of neurotransmitter its effect might be excitatory or inhibitory on the target cell. Motor neurons carry nerve impulses from the central nervous system to the muscles of the body. ACh is released from the nerve ending into neuro-muscular junction triggered by nerve action potential. If sufficient amount of ACh is released into the neuro-muscular junction, muscle action potentials arise and propagate into opposite directions toward the tendons of the muscle (Fig. 2.3).

The action potential in human nerve-cells reaches an amplitude of nearly 90 millivolts and has a duration of about 3 - 4 milliseconds. Although the resulted

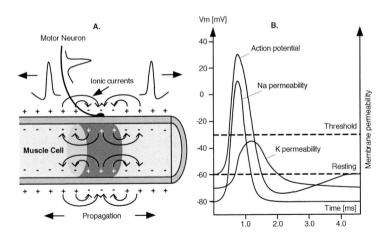

Figure 2.3: ***A.*** *A muscle action potential triggered by nerve action potential.*
B. *Nerve Action Potential.*

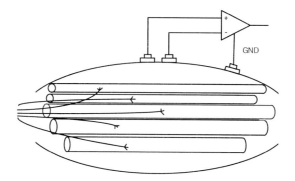

Figure 2.4: Surface recording of muscle signals.

frequency of action potentials is less than 300 Hz a high frequency recording is frequently used to be able to better resolve several action potentials from neighbor cells (for example: Evoked Potential Recording).

2.2   Biopotential Recording System

Most common electrophysiologic measurements take the potential change in the extracellular medium of bio cells (muscle and nerve cells) into consideration. Therefore a basic understanding of the extracellular potential is important (see appendix A.1). The voltage measured externally is considered as a potential difference between two or more spots on the body tissue. A sum of extracellular potentials from several nerve/muscle cells is recorded at every spot using an appropriate bio electrode. The placement of the bio electrodes relatively to the source of bio-potential and the distance between the electrode contacts plays an important role in the shape of the recorded signal (more information for this purpose is provided in section 2.2.1). The electrode is connected to a proper electrical instrumentation system for signal conditioning and further processing (Fig. 2.4).

Common physiological measurements such as Electrocardiograph (ECG), Electroencephalograph (EEG), and Electromyograph (EMG) provide physicians with important information about the function of an organ in the body of a patient. Table 2.1 presents an overview of common physiological surface recordings [3]. Bio-signal properties show big similarities in bandwidth and dynamic range. They conclude altogether a low frequency range and small signal amplitudes. This is an important specification value for bio recording systems. It defines mainly the bandwidth of analogue front-end and the gain factors of amplifier devices.

Table 2.1: Bio-Signal Measurements

Measurement	Bio-Signal Characteristics	Acquisition
Electrocardiogram (ECG)	bandwidth: 0.05 - 150 Hz Dynamic Range: 1-10 mV	Surface electrodes
Electroencephalogram (EEG)	bandwidth: 0.5 - 100 Hz Dynamic Range: 2-100 μV	Surface electrodes
Evoked potentials (EP)	bandwidth: 2 Hz-3 kHz Dynamic Range: 0.1-20 μV	Surface electrodes
Electromyogram (EMG)		
Single-fiber (SFEMG)	bandwidth: 500 Hz-10 kHz Dynamic Range: 1-10 μV	Needle electrodes
Motor unit AP (MUAP)	bandwidth: 5 Hz-10 kHz Dynamic Range: 100μV-2 mV	Needle electrodes
Surface EMG (SEMG)		
Skeletal muscle	bandwidth: 2-500 Hz Dynamic Range: 50μV-5 mV	Surface electrodes
Smooth muscle	bandwidth: 0.01-1 Hz	Surface electrodes

Most bio-signal measurement systems consist of three main components:

- **Bio Sensor:** It represents the interface between the source of activity and the analogue front-end of an electronic circuit for bio-signal conditioning. Bio sensors collect the bio potential either from the body surface such as surface electrodes or from the inside of the body (invasive electrodes) such as needle electrodes.

- **Conditioning circuit and signal transmission:** It concludes analogue and digital electric circuits for bio-signal conditioning connected to the bio electrodes, for example: Bio-amplifier, filter, analog-to-digital converter (ADC). An application-specific integrated circuit (ASIC) is best used for this purpose. A control unit is then used to transmit the digitized bio-signals to the processing and display units. A microprocessor and wireless data transmission chips may be used for this purpose.

- **Signal processing and representation:** Further processing of the received signals takes place in digital form if needed. In this part of the system digital filtering and algorithms for information extraction out of the raw bio-signals may be used. Furthermore, the signals may be displayed on a screen to give the physician a close view of the shape of the recorded signals.

A close look on the components of bio recording systems with focus on implantable applications is depicted in the following sections.

2.2.1   Electrodes

The transfer of signals inside the body involves ions as charge carriers. On the other hand, conventional electronic systems utilize electrons for the same purpose. Collecting bio-signals from biological tissue or stimulating excitable tissues requires an interaction between biological and electronic systems. Electrodes represent a part of an electronic system, where the ionic currents are transduced into an electron current. It consists of electrical conductors placed in contact to the body to pick up bio-electrical activity. At the interface between an electrode and an ionic fluid (electrode-electrolyte interface) a process (chemical reaction) must take place for transmission of charge between the electrode and the ionic solution. If the solution contains the same type of ions as in the electrode a concentration gradient emerges locally in the electrode-electrolyte interface which in turn leads to a potential difference between the electrode itself and the electrolyte. This potential difference is known as the half-cell potential (Fig. 2.5). It is very important for designing the electrical circuitry for signal acquisition. The values of the half-cell potential can reach 200 mV or a voltage high enough to drive a dc-coupled differential amplifier in the conditioning circuitry into saturation [3].

The half-cell potential is altered due to temperature change and ionic activity in the electrolyte [15]. As a result the half-cell potential receives an *overpotential*. The *overpotential* is subdivided into three components:

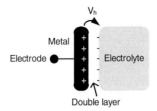

Figure 2.5: The double layer of charge in the electrode-electrolyte interface (Helmholtz 1879). V_h is the half-cell potential.

- *The ohmic overpotential* depicts the resistance of the electrolyte itself as a current passes through.

- *The concentration overpotential* describes the effect of changes in the ion distribution close to the electrode-electrolyte interface when a current is established.

- *The activation overpotential* represents the activation energy which controls the oxidation-reduction process of metal atoms and their movement between metal and electrolyte. Due to the direction of current, the atoms must overcome a certain energy barrier for moving into electrolyte or back to the metal. The energy barrier differs according to the direction of current. The activation potential mainly shows this difference in terms of voltage component across the electrode-electrolyte interface.

Fig. 2.6 shows the process of charge transfer over the electrode electrolyte interface. As a result of oxidation-reduction reactions one metal atom M is oxidized out of the metal and gives off one free electron (e^-) towards the metal. On the other hand, one cation M^+ becomes neutral when moving to the metal and getting one free electron. The chemical reaction at the electrode-electrolyte interface is described as follows:

$$C \rightleftharpoons C^{n+} + ne^- \qquad (2.2)$$

$$A^{m-} \rightleftharpoons A + me^- \qquad (2.3)$$

where n and m contains the valence of C and A respectively.

Theoretically, biopotential recording electrodes can be divided into two types according to the chemical reaction at the electrode-electrolyte interface due to current flow:

- **Polarized electrodes:** electrode behaves as a capacitor as no charge passes the interface but a displacement current.

- **Nonpolarized electrodes:** current freely passes the interface without the need of energy to cross the interface. Therefore no *overpotentials* occur.

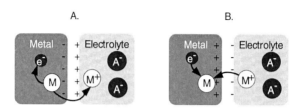

Figure 2.6: Charge transfer over the electrode-electrolyte interface.

Real electrodes show properties from both types. Some electrodes are very close to the nonpolarized property such as silver/silver chloride electrodes and others are closer to perfectly polarized ones such as those made of noble metals (platinum electrodes). The properties of an electrode concerning polarization are very important for the designer. Polarized electrodes show a strong capacitive characteristic. As a result they are very sensitive to motion and have a reduced performance [3].

Electrical information about the electrode-electrolyte interface is essential for stimulating and recording circuit design. The first modeling of it was done by Helmholtz [16] in 1879 with a proposal of the double layer of ions with opposite charges at the metallic surface and the electrolyte (Fig. 2.5). The most important aspect of this concept is the existence of a capacitance and a resistance in any equivalent model for the interface based on the idea that a charged capacitor exists in the double layer and a direct current is transduced by the interface (voltaic cell). The capacitance might reach values in the $\mu F/cm^2$ range as the distance between the charges in the double layer is very small ($\sim 10^{-10}$m) [17]. At 1899-1901 Warburg proposed a relationship between the polarization capacitance of the double layer and the frequency ($C_w = k/f^{0.5}$) for an infinitely low current density [18][19]. In 1968, Geddes and Baker introduced two equivalent circuit models for the electrode-electrolyte interface [20]. The first model concludes a shunt circuit of the Warburg capacitance (C_W) to a Faradic resistance (R_f) representing the ohmic current property of the interface (Fig. 2.7 A). In the second model the Faradic

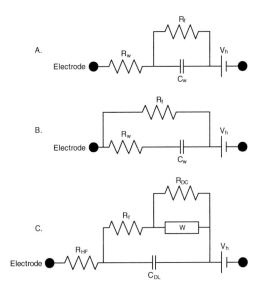

Figure 2.7: The equivalent circuits of the electrode-electrolyte interface (A and B: Geddes and Baker 1968, C: Randles 1951). V_h is the half-cell potential, R_w and C_w represent the Warburg element, and R_f is a Faradic resistance.

resistance (R_f) is shunted to the complete Warburg element (Fig. 2.7 B). A model for better interpretation of the impedance spectrum of the electrode-electrolyte interface was proposed by Randles 1951 [21]. Fig. 2.7.C shows an equivalent circuit of Randles, where the electrolyte resistance is represented by R_{HF}, C_{DL} is the double layer capacitance, R_f is a faradic resistance, W represents the Warburg element, and R_{DC} is a correction resistance. The Randles model is an empirical one which delivers good approximation to measurement curves.

The path between the source of bio activity (nerve or muscle cell) and the electronic circuitry consists of many impedance components. Fig. 2.8 shows a simplified equivalent circuit of some of the impedance components. The skin exhibits the highest impedance of body tissue of the signal path to the electrode. Furthermore, it possesses a low pass filter characteristic. To reduce the impedance of the skin (mainly contributed by the epidermis "dead outer skin layer") the epidermis is usually removed in the preparation phase preceding to a surface recording [22].

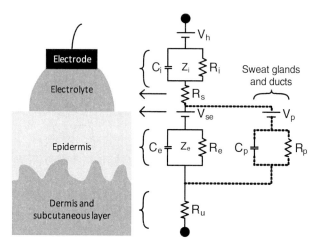

Figure 2.8: The equivalent circuit of the electrode-skin interface including the skin layers.

Many types of electrodes have been developed for different biomedical recordings. This section presents a brief description of the frequently used electrodes.

Surface Electrodes

Surface electrodes are placed on the body surface for non-invasive biopotential recording. They are used for short term (diagnostic recording of ECG) as well as for long term recordings (cardiac monitoring) [3]. For the later, nonpolarizable electrodes are used to maintain interface stability and hence better long term recording performance. Several materials are used to build surface electrodes such as: Silver/Silver Chloride, Gold etc.
Fig. 2.9 shows the common used surface electrodes.

Indwelling Electrodes

Indwelling electrodes are used for invasive application. They penetrate the skin and are placed close to the source of biopotential. Thus they represent a localized pickup of activity and hence possess a higher selectivity from surface electrodes. Biopotentials from single muscle or nerve cells can be recorded with this kind of electrodes. There are several types of indwelling electrodes according to their application. The

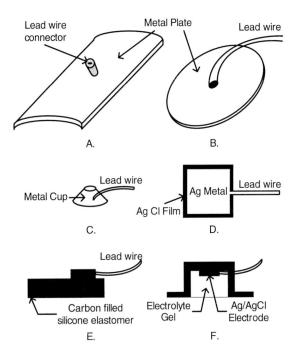

Figure 2.9: Commonly used surface electrodes [3]: A. and B. metal plate electrodes (commonly used in ECG recording), C. metal cup electrode (Gold electrode with gel inside. Commonly used in EEG recording.), D. silver/silver chloride surface layer, E. carbon-filled elastomer dry electrode, F. recessed electrode (reduction of movement artifact).

common types are as follows:

- **Fine wire electrodes:** The fine wire electrodes (commonly used material is stainless steel with a diameter of 25 to 125 μm [15]) are manufactured within a needle as shown in Fig. 2.10.d. With the help of the needle the fine wire is inserted into the body to the target location. The needle is then drawn out leaving the fine wire inside the body. Fine wire electrodes are more flexible for percutaneous recording than hard needle electrodes. Therefore they are used for long term implantation and chronic recordings.

- **Needle electrodes:** It consists of a needle (commonly used material is stainless steel) with sharp ending (tip of the needle). According to recording art (monopolar or bipolar recording) there might be one (Fig. 2.10.b) electrode or two (Fig. 2.10.c) at the tip of the needle. The diameter of the electrode is in the micrometer range (25 μm). It is used for example to detect the activity of one or more muscle fibers of a motor unit. As a result the properties of the motor unit can be monitored (firing rate, recruitment, etc.).

- **Micro electrodes:** Micro electrodes or micropipets are mainly used to study the behavior of membrane channels in excitable biological cells (muscle and nerve cells). Microelectrodes consist of a glass shaft filled with metal internally. Whereas micropipets contain electrolyte material inside the glass shaft and a metal electrode connected to conditioning circuitry.

- **Electrode arrays:** This type of electrodes contains a dense array sensor for recording bio activity of single nerve cells as well as for stimulating single nerve cells. The electrode array can stimulate neurons selectively by directing the electrical field with using several electrodes. The application field of this kind of electrodes is growing rapidly as technology possibilities of manufacturing and integrating of electronic circuits and electrodes are developing. The most common fields of application for array electrodes are: visual neuroprosthesis [23], optical stimulation of neurons [24], implants for people suffering from neurological disorders [25], Utah electrode [26], the Michigan array [27], or the new NeuroProbes arrays for cerebral application [28].

- **Implantable myo-electrodes:** In the control of myogen controlled neuro- or myoprostheses implantable electrodes are used to supply the system with multiple muscle control signals. Intramuscular and epimysial electrodes have been implanted with a neuromuscular stimulator for the use in neuroprostheses [29] [30] [31].

Thin film polymid based electrodes have been developed for recording of intramuscular signals [32]. An investigation of the usability of implantable thin film electrodes has been presented [33] [34]. Although the flexible film-based electrode shows good mechanical stability, the connector to such electrodes reveals a weakness for long-term implantation.

Intramuscular electrodes (IMES) have been used in an encapsulated system for the recording of muscle activity to control a hand prosthetic [35]. This technique has the main advantage of eliminating the problems associated with passing wires percutaneously, such as infections, cable movement artifacts, and wire breakage. Furthermore the long-term stability has been approved for over two years in rhesus monkeys [36]. A sampling rate of 1.26 kHz for each sensor is used. Internal EMG recording is believed to deliver high frequency components of the EMG signal. This might offer an opportunity for better study of the internal EMG signal components regarding prosthetic control purposes. The low sampling rate of the IMES might cause aliasing in the recorded EMG signal.

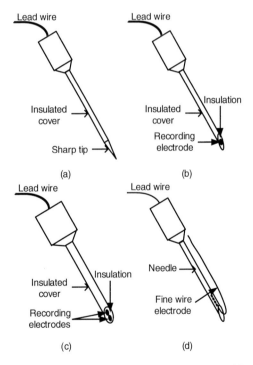

Figure 2.10: Fine wire and needle electrodes: (a) needle electrode, (b) monopolar needle electrode, (c) bipolar needle electrode, (d) fine wire electrode.

Electrode Configuration

Electrode configuration defines the number of biopotential recording sites and their position as well as the contact of the sites to the electronic circuit (biopotential amplifier). The most common electrode configurations are:

- **monopolar recording:** a monopolar recording measures the potential difference between an active region and an inactive one (Fig. 2.11.a). The placement of the monopolar electrode mainly affects the shape of the recorded wave. The properties of monopolar recording are as follows:

  1. The recorded signal consists of the amplified difference between a source signal (S in Fig. 2.11.a) and a reference signal on an inactive region $Gain * (S - Ref + noise)$.

  2. Different noise waves couple into the recorded signal. The contaminated noise remains in there throughout processing.

  3. In a multi-channel recording system there is a higher risk of losing a shared reference electrode. The electrode loss would in that case affect all the channels of the recording system.

- **bipolar recording:** a bipolar recording measures the potential difference between two active regions (Fig. 2.11.b). The placement and the distance between electrodes in a bipolar recording are more determined for the recording wave. The properties of bipolar recording are as follows:

  1. The recorded signal is basically the amplified difference between neighboring electrodes on an active region $Gain*[(S_1+noise)-(S_2+noise)] \approx Gain*(S_1-S_2)$. Bipolar recording takes advantage of high common mode rejection ratio (CMRR) of bio amplifiers in order to eliminate common mode noise.

  2. The differential amplification circuit is less prone to interference from adjacent and deeper muscles.

  3. Better symmetry properties for the analog amplifier circuit.

The bipolar recording provides a more selective approach of biomedical recording. Only the signals originated from the muscle of interest are gathered. Therefore the bipolar acquisition method could be suitable for providing independent control signals to a prosthetic system with many degrees of freedom.

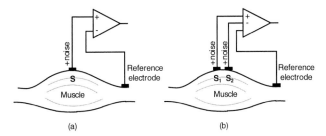

Figure 2.11: Most common electrode configurations: (a) monopolar recording, (b) bipolar recording.

Electrode noise

One of the most important parameters characterizing a biopotential recording system is the ratio of energy of bio-signal divided by the energy of noise signal "Signal-to-noise ratio" (SNR). There are two main sources for noise contributed to the signal of interest: biological artifacts and electronic or electromagnetic noise. Biological artifacts contains biological signals of no interest overlapping the signal of interest (muscle artifacts into EEG recording). Electronic noise includes the noise components originated in the electronic circuitry of the recording system (thermal noise or flicker noise in amplifiers) or from the surrounding area of the recording such as line noise (also called 50 Hz noise). The electrode interface to the body depicts the first noise contributor in the recording system. The equivalent circuit of that interface shows capacitive and resistive components which in turn contribute with a noise factor to the recorded signal. The electrode interface to the body delivers thermal noise which can be calculated by the Nyquist equation [37]:

$$V_{th}^2 = 4kTR \cdot \Delta f \quad [V^2] \quad (2.4)$$

where V_{th}^2 is the thermal noise power, k is the Boltzmann constant ($1.38 \cdot 10^{-23} JK^{-1}$), T is the temperature (in Kelvin), R is the resistance (in ohms), and Δf is the frequency bandwidth (in Hz). A noise magnitude of $10 - 60$ μV_{PP} is measured from the body surface [38]. Noise magnitude at the electrode seems to be inversely

proportional to the square root of the area of the electrode on the skin ($noise \propto \frac{1}{\sqrt{A}}$)
[39]. Typical noise level of the Michigan electrode is 15 μV$_{RMS}$ ($20-40$ μV$_{PP}$) over
the normal unit recording bandwidth [40]. The impedance lies between $0.5-3$ MΩ
dependent on the size of the electrode.

2.2.2   Conditioning Circuits and Signal Transmission

This section will describe the properties and characteristics of electronic circuits used
for biopotential signal acquisition in implantable systems. These include biopotential
amplifiers, analogue filters, analogue to digital converters, wireless data and energy
transmission for implantable devices.

2.2.2.1   Biopotential Amplifiers

The amplifiers used for surface biopotential-recordings (EMG, EEG, ECG etc.) must
meet similar specifications considering noise, signal bandwidth, and gain factors.
The amplifiers for implantable applications have moreover a stringent specification
for power consumption (low-power). Therefore a trade-off between power consumption and noise is mostly done when designing analogue circuitry. In implantable
applications the area must be considered in the trade-off as well. This section will
summarize the important properties of biopotential recording amplifier.

Input impedance

The path from the biopotential source to the amplifier circuitry passes several stages.
Fig. 2.12 shows the main impedance components in the path. Biopotentials nor-
mally have low values in the micro Volt range (Table 2.1). A high signal-to-noise
ratio (SNR) is very important when acquiring biological activity. There are two
ways to increase SNR of bio recorded signals: reducing the electrode impedance
and increasing the input impedance of the amplifier. Small electrodes for internal
recording of nerve or muscle activity possess a small area and hence a high impedance
(several mega ohms [40]). Therefore the input impedance of the amplifier must be
greater than 100 MΩ for a 100 to one relation of the voltage drop between the input
impedance of the amplifier and that of the electrode interface neglecting the small
tissue impedance.

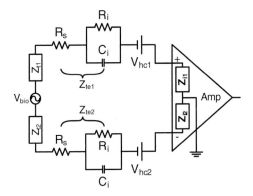

Figure 2.12: Equivalent circuit of the impedance at the input of bio-signal recording amplifiers. V_{bio} is the source bio potential, $Z_{t1,2}$ is the tissue impedance, $Z_{te1,2}$ is the impedance of the electrode-tissue interface, $V_{hc1,2}$ is the half-cell potential, $Z_{i1,2}$ is the amplifiers input impedance.

Gain and Bandwidth

Biopotential amplifiers typically have a gain of 60 dB ($=$ x1000) [41] to increase the resolution of the small recorded biopotentials. The magnitude of bio-signals depends among other things on the placement of the electrodes and their inter-electrode distance. Thus a programmable gain is usually needed for better resolution of the non-predictable small amplitudes at the amplifier's input. Their bandwidth spread from sub Hertz region to several tens of kilo Hertz (Table 2.1). Bio recording amplifiers are designed to a limited bandwidth. The purpose of band limited amplifiers is to eliminate unwanted signals such as electrode half-cell potentials at dc-level and high frequency interference. Furthermore, high bandwidth option is used for testing the low-pass characteristic of the skin by comparing invasive EMG with parallel recorded surface EMG.

Noise

The noise exists in the nature of the electronic elements which builds the amplifier circuitry (active components: transistors, passive components: resistances, capacitors etc.). Input noise generator models are frequently used for noise analyses in amplifiers. The main idea behind it is basically that the internal noise sources of a

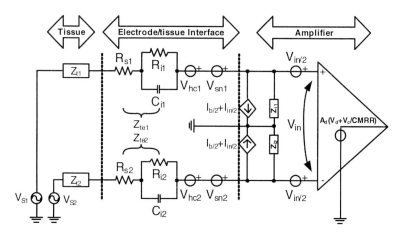

Figure 2.13: Equivalent circuit of the input stage of a bio-signal recording system including noise sources. $Z_{t1,2}$ is the tissue impedance, $Z_{e1,2}$ is the tissue-electrode impedance, $V_{hc1,2}$ is the half-cell potential, $V_{sn1,2}$ is the tissue-electrode noise plus electrode thermal noise, $I_{b/2}$ is the bias current of amplifiers input stage, $Z_{i1,2}$ is the input impedance of the amplifier, V_{in} is the input referred voltage noise of the amplifier, $I_{in1,2}$ is the input referred current noise, A_d is the differential gain, V_d is the differential input voltage, V_c is the common mode input voltage, and CMRR is the common mode rejection ratio.

linear twoport can be represented by a voltage source acting in series with the input voltage (input-referred differential voltage noise, V_{in}[1]) and a current source flowing in parallel with the input current (current noise in the inverting and non-inverting inputs of the amplifier, I_{in}[2]) [42]. Fig. 2.13 modifies the equivalent circuit from Fig. 2.12 by including the noise sources of the electrode-tissue interface and those of the amplifier.

Noise powers in a recording system are additive whereas noise voltages add in a root sum of squares assuming they are uncorrelated with each other. As a result of that, biggest noise values dominate the overall noise of a system. The noise of the input stage of the bio recording system constitutes of intrinsic noise of the amplifier, the electrode thermal noise (equation 2.4), and the electrode contact noise (generated by ionic exchanges at the tissue-electrode interface [43]). It can be subsumed as

---

[1]V_{in} is determined by shorting the input to ground.

[2]I_{in} is determined by floating the input terminal.

follows:

$$V_{\text{noise,RMS,total}} = \sqrt{V_{\text{noise,amp}}^2 + V_{\text{thermal,RMS}}^2 + V_{\text{contact,RMS}}^2} \qquad [V_{\text{rms}}] \qquad (2.5)$$

Lowering the intrinsic noise of the bio amplifier is very important for present and for future applications. Thus much research is taking place in bio electrode development to reduce the impedance [44] and hence to improve the performance and electro-chemical properties of bio interfaces [45].

Power Consumption

Implantable devices may have two sources of power: a battery (such as battery pow-ered pacemakers [46]) or wireless power transmission [47], [48]. Batteries consume a certain volume in an implantable device and might get empty and hence exhibit some risks during replacement (infection risk etc.). Therefore wireless transmission de-picts the better long term solution for implantable devices in terms of risk reduction and volume occupation. However, energy losses almost always accompany wireless transmission systems. Therefore implantable systems are supposed to consume low power.

One approach to decrease power consumption is the use of Low V_{dd} technologies, although they are more expensive and show more noisy transistor characteristics. But with a low-noise amplifier topology and an application specific structure, it is possible to accomplish power efficient acquisition systems.

Bias currents in amplifier design correlate with the power consumption of the amplifier. Low currents are desirable for low power application. Thus MOS transis-tors in the sub-threshold region are best suited for low power applications as they deliver high g_m values and hence it is possible to achieve high amplification factors with reasonable noise values.

Common Mode Rejection Ratio (CMRR)

Many biopotential recordings utilizes bipolar recording methods. When electrodes are placed close to each other the external coupled noise and other unwanted signals couple almost the same on both amplifier inputs. The common mode rejection ratio is a value which depicts the ability of the amplifier to reject unwanted common signals on the two inputs of a differential amplifier. It is calculated as follows:

$$\text{CMRR}_{\text{dB}} = 20 \log \frac{A_d}{A_c} \qquad (2.6)$$

where A_d is the differential gain of the amplifier (different signals at the input) and A_c is the common mode gain (identical signals at the input).

A monopolar configuration doesn't take advantage of the CMRR of bio-amplifiers. Therefore, the CMRR value is not important for monopolar recordings.

Input Coupling

Two methods exist to couple the input signal on the bio amplifier:

- **DC-coupling:** Biopotentials are connected directly or through resistors to the input of the amplifier which allows DC-signals to be recorded and further amplified by the amplifier.

- **AC-coupling:** DC-content of the input signal is high-pass filtered using a capacitor in series to the input path. A high pass characteristic is accomplished herewith.

Section 2.2.1 mentioned the so-called half-cell potential at the electrode-electrolyte interface. This potential is a dc-potential and exists at the input of a recording amplifier. The dc-potential might change according to movement and disruption of the electrode-electrolyte interface. Therefore an ac-coupling method is advantageous to overcome the dc-potentials at the input terminal of an amplifier. This is indeed a challenging aspect in biomedical system design. On the one hand low signal frequencies are important for the signal analysis and on the other hand a high pass filter is implemented using the ac-coupling method. The solution is to push the cutoff frequency, f_{3dB}, of the high pass filter as low as possible. In return high RC value is needed. As capacitors are area consuming elements, the best solution is to increase the resistance using diode connected transistors [49].

2.2.2.2   Active Filters

Active filters are designed as analogue circuits used to remove unwanted frequency components or frequency range from a recorded signal while passing the signals at a certain frequency range. Unwanted signals in a biopotential recording have two sources:

- **electromagnetic noise:** this category is basically made of electrical signals from electrical devices such as:

1. high frequency electromagnetic waves coupled from electric devices in the surrounding,

2. low frequency movement artifacts such as cable or electrode movement, and

3. 50 Hz line noise etc.

- **biological artifacts:** the noise from a biological origin might be:

1. muscle activity in an EEG recording, or

2. dc-potentials such as the half-cell potential from the electrode-tissue interface etc.

A combination of high-pass and low-pass filters can be used at the analogue front end of a biopotential recording system (Fig. 2.14) in order to remove unwanted low frequency and high frequency noise signal ranges respectively. The noise frequencies that exist in the range of recorded biopotential, such as 50 Hz line noise, are usually not filtered in order not to remove useful bio-signal components. If needed they might be digitally filtered using a band-stop filter (50 Hz band-stop filter).

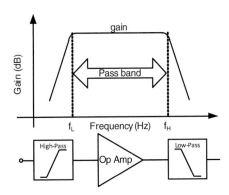

Figure 2.14: Low-pass and high-pass filter characteristic at the analogue front end.

Furthermore, analogue filters are used to avoid aliasing in the converted digital signal. Some system-theoretical fundamentals are needed to illustrate the benefits of anti-aliasing filter. Continuous analogue signals are converted into a series of discrete values with defined sampling frequency $f_a = 1/T_a$ (Fig. 2.15). The discrete

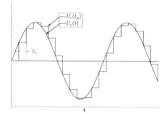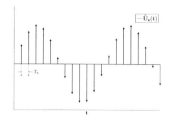

Figure 2.15: Analog input signal $U_e(t)$ with its discrete samples $U_e(t_\mu)$ and an illustration of the analogue input signal using a Dirac series.

signal fits the continuous signal better when a higher sampling frequency is used. It can be interpreted as a series of Dirac-pulses as follows:

$$\tilde{U}_e(t) = \sum_{\mu=0}^{\infty} U_e(t_\mu) T_a \delta(t - t_\mu) \tag{2.7}$$

By using the Fourier transformation it is possible to calculate the spectrum of the discrete signal to extract the information content of the Dirac series:

$$\tilde{X}(jf) = T_a \sum_{\mu=0}^{\infty} U_e(\mu T_a) e^{-2\pi j \mu f / f_a} \tag{2.8}$$

where the sampling frequency f_a is the period of the spectrum function. The spectrum of equation 2.8 is identical to the spectrum of the original continuous signal function $|X(jf)|$ in the range $-\frac{1}{2}f_a \leq f \leq \frac{1}{2}f_a$ (Fig. 2.16). The sampling frequency must be high enough though to prevent frequency overlap in the spectrum function[3]:

$$f_a > 2f_{max} \tag{2.9}$$

The low-pass filter must be designed to pass on the spectral components till f_{max} and to reach an infinite attenuation already in $\frac{1}{2}f_a$ [50]. It is possible to reconstruct a band limited analogue signal from the sampled discrete signal only if the condition of the sampling theorem is fulfilled ($f_a \geq 2f_{max}$). For that purpose, a series of Dirac-impulses from the sampled values are to be constructed and fed through a low pass

---

[3]This is the sampling theorem according to Nyquist and Shannon.

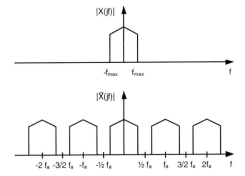

Figure 2.16: The spectrum of the analog input signal (top) and that of the sampled signal (bottom).

filter with cutoff frequency $f_g = f_{max}$. Aliasing occurs when the sampling frequency is lower than described in the sampling theorem. The low pass filter will not be able to attenuate some spectral components with a frequency difference of $f_a - f < f_{max}$. This results in a beat frequency at the output (Fig. 2.17).

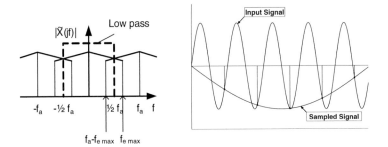

Figure 2.17: Overlap of spectral components (left) and beat frequency (right) when low sampling frequency is used.

2.2.2.3   Analogue to Digital Converters

The ADC is considered to be the last component of a recording channel. The amplified analogue signals at the input of an ADC are converted into digital words for further signal processing. There is a variety of ADCs for digitizing analogue signals. A well defined specification of the used system would certainly facilitate the choice of a certain ADC for the target application. The characteristics of an ADC can be summarized as follows:

- **Input range:** The input range specifies the voltages at the input of the ADC. Those voltages are restricted with a maximum value called the voltage reference (V_{ref}) of the converter.

- **Resolution:** The resolution of the ADC describes the relationship between the digital value at the output and its related input voltage at the input taking the voltage reference into consideration. The smallest represented digital term, also called LSB, can be calculated as: $LSB = \frac{V_{dd}}{2^n}$, where n is equal to the number of bits at the output of the ADC and V_{dd} is the voltage reference of the ADC. For example the size of the step of a 10 bit ADC with a 1.2 V reference voltage is given by:
$$\text{LSB} = \frac{\text{V}_{\text{ref}}}{2^{10}} = \frac{1.2\text{V}}{1024} = 1.17\text{mV}$$

- **Digital coding:** The representation (of one's or two's complements etc.) of the sampled digital values is necessary for the processing units to be able to interpret the digital values.

- **Data rate:** The sampling frequency of the ADC defines the data rate at the output of the electronic system. The frequency of the source input signals and the oversampling defines the data rate of the ADC.

- **Quantization error:** The quantization error, Q_e, is the difference between the analogue input value and the (staircase) digital value at the output of the ADC. It is calculated as follows [51]:
$$\text{Q}_e = \text{v}_{\text{IN}} - \text{V}_{\text{staircase}} \tag{2.10}$$
where v_{IN} is the input analogue voltage and $V_{staircase}$ is the ADC output voltage which is calculated as:
$$\text{V}_{\text{staircase}} = \text{D}\frac{\text{V}_{\text{ref}}}{2^{\text{N}}} = \text{DV}_{\text{LSB}} \tag{2.11}$$

where D is the digital value at the output and V_{LSB} is the resolution voltage of 1 LSB.

- **Static performance:** Several terms exist to specify the performance of an ADC such as offset error, gain error, integral nonlinearity (INL), and differential nonlinearity (DNL). The offset error is a horizontal shift in the ideal characteristic of the ADC (horizontal shift of the ideal 3-bit signal in Fig. 2.18). The gain error describes the difference of the ideal slope (ideal characteristic) and the actual slope of the output of the ADC. It could be calibrated out by multiplying the final conversion by a constant value [52] (this would increase the processing overhead in the system). Integral nonlinearity (INL) is the maximum difference between the ideal characteristic signal and the actual converted signal measured vertically in percent or LSBs [53]. Differential nonlinearity (DNL) measures the separation between adjacent digital codes at each vertical step (each transition) in percent or LSBs. It is given as follows:

$$\text{DNL} = (\text{D}_{\text{cx}} - 1)\text{LSBs} \tag{2.12}$$

where D_{cx} is the size of the actual vertical step in LSBs.

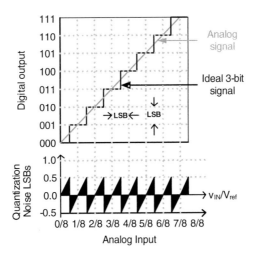

Figure 2.18: *Input-output characteristics of an ideal 3-bit ADC.*

- **AC characteristics:** Signal to noise ratio (SNR), effective resolution (ER), signal to noise and distortion ratio (SINAD), and effective number of bits (ENOB), conclude information about ADC repeatability [52]. SNR represents the ratio between the signal of interest and quantization noise generated in the converter. It is calculated as the ratio between the largest RMS input signal into the ADC over the RMS value of the noise:

$$\text{SNR} = 20 \log(\text{V}_{\text{in\_max\_rms}}/\text{V}_{\text{noise\_rms}}) \tag{2.13}$$

The maximum input signal of an ADC with a full scale peak-to-peak sinus wave amplitude V_{ref} is calculated as:

$$\text{V}_{\text{in\_max\_rms}} = \frac{\text{V}_{\text{ref}}}{2\sqrt{2}} = \frac{2^{\text{N}}(\text{V}_{\text{LSB}})}{2\sqrt{2}} \tag{2.14}$$

The quantization error has a uniform distribution and is equally likely (Fig. 2.18). Its uniform distribution ranges from $-0.5LSB$ to $+0.5LSB$. The RMS quantization error is then calculated as:

$$\text{V}_{\text{noise,rms}} = \sqrt{\frac{1}{\text{V}_{\text{LSB}}} \int_{-0.5\text{V}_{\text{LSB}}}^{0.5\text{V}_{\text{LSB}}} \text{V}^2 \, d\text{V}} = \sqrt{\frac{1}{\text{V}_{\text{LSB}}} \left[\frac{\text{V}^3}{3}\right]_{-0.5\text{V}_{\text{LSB}}}^{0.5\text{V}_{\text{LSB}}}} = \frac{\text{V}_{\text{LSB}}}{\sqrt{12}} \tag{2.15}$$

The ideal SNR value of an ADC is then calculated as:

$$
\begin{aligned}
SNR_{ideal} &= 20 \log \frac{\frac{2^N(V_{LSB})}{2\sqrt{2}}}{\frac{V_{LSB}}{\sqrt{12}}} \\
&= 20N \log(2) + 20 \log(\sqrt{12}) - 20 \log(2\sqrt{2}) \\
&= 6.02N + 1.76
\end{aligned} \tag{2.16}
$$

The effective resolution (ER) is calculated as follows:

$$\text{ER} = \text{N} - \log_2(\sigma) \tag{2.17}$$

where N is the bit resolution of the ADC and σ is the standard deviation of the output signal. SINAD is defined as:

$$\text{SINAD} = \frac{\text{P}_{\text{signal}} + \text{P}_{\text{noise}} + \text{P}_{\text{distortion}}}{\text{P}_{\text{noise}} + \text{P}_{\text{distortion}}} \tag{2.18}$$

where P represents the power of the signal, noise and distortion components in the measured signal. From the SINAD we can calculate the ENOB as follows:

$$\text{ENOB} = \frac{\text{SINAD} - 1.76}{6.02} \tag{2.19}$$

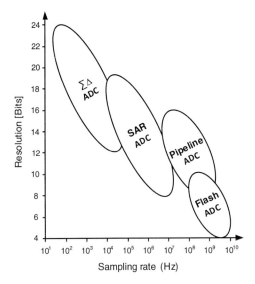

Figure 2.19: ADC types and their characteristics concerning sampling rate and bit resolution.

Analogue-to-Digital converters are usually characterized by their Bit-resolution, N, and by their sampling frequency, f_s. Several types of ADC architectures exist. The main four types of ADCs are: oversampled ADCs ($\Sigma\Delta$ ADC), successive approximation register ADC (SAR), pipeline ADC, and flash ADCs. They show differences in the resolution and the speed of conversion as shown in Fig. 2.19. Bio-signals have a low frequency range. Therefore the used ADCs for biopotential recordings are usually the SAR and $\Sigma\Delta$ architectures as they satisfy several multiples of the sampling frequency for bio-signals according to Nyquist (equation 2.9). The following paragraphs provide a brief description of the main used ADC architectures for biopotential recordings.

The Oversampled Converters

Oversampling converters sample the analogue signal at a rate much higher than the bandwidth of the input signal. Despite their high sampling rate (oversampling), the throughput of such an ADC is lower than the Nyquist rate ADCs because of

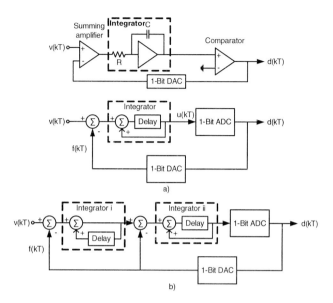

Figure 2.20: Sigma-delta modulator with a first order modulator a), and a second order modulator b).

the time-consuming sampling process. They can reach high resolution values due to the utilization of digital processing techniques instead of complicated and precise analogue building components.

Fig. 2.20.a illustrates a first order oversampled ADC, the sigma-delta modulator. Herein an integrator and a 1-bit ADC are used in the signal direction, whereas a 1-bit DAC is used in the feedback path. T is the sampling time $(T = 1/f_s)$, and k is an integer value (k = 1, 2, 3...). The 1-bit ADC delivers a digital '1' or digital '0' at the output, which can be realized using a simple comparator. The 1-bit DAC is used to feedback a V_{ref} or $-V_{ref}$ to be summed up with the input signal. The characteristics of the sigma-delta modulator can be calculated as follows:

$$u(kT) = v(kT - T) - f(kT - T) + u(kT - T) \qquad (2.20)$$

The quantization error of the 1-bit ADC is defined as:

$$Q_e(kT) = d(kT) - u(kT) \qquad (2.21)$$

The digital output can then be calculated as:

$$d(kT) = Q_e(kT) + v(kT - T) - f(kT - T) + u(kT - T) \qquad (2.22)$$

The 1-bit DAC can be realized using switches connected to $-V_{ref}$ and to V_{ref} which idealizes the behavior of the DAC. This equalizes the output value to the feedback value ($d(kT) = f(kT)$). As a result:

$$d(kT) = v(kT - T) + Q_e(kT) - Q_e(kT - T) \qquad (2.23)$$

Equation 2.23 indicates that the output signal is delayed by one period. Furthermore, the quantization noise, $Q_e(kT)$, is mostly canceled out due to the subtraction of noise effect from last period. A higher order sigma-delta modulator can be designed to improve the noise shape and increase resolution of the ADC if combined with higher oversampling ratio (Fig. 2.20.b). An increase of 9 dB in dynamic range (~ 1.5 bit in resolution) can be reached with every doubling of the oversampling ratio using a first order modulator [51].

Successive Approximation Register Converters

The SAR converter is frequently used in today's applications. One can reach high resolution and high sampling rates using this architecture. It consists basically of three components: a comparator for comparing quantization levels, a DAC to provide the compare voltage, and a register block including the successive approximation register for the binary search of the input voltage. Fig. 2.21 illustrate a block diagram of the successive approximation register (SAR) ADC. The SAR utilizes a binary search for the input voltage by a successive shift of a '1' through its shift register and saving that '1' into the SAR register in case the value is greater than the input voltage on the comparator. The DAC in this architecture is a sensitive component as an error in the digital to analogue conversion step would lead to a fatal error in the output register due to the search in a wrong binary section.

The most frequent way to implement SAR converters is by using a binary weighted capacitor array as its DAC (also called the charge-redistribution successive approximation ADC). Fig. 2.22 depicts a block diagram of the converter. The basic functionality of this converter is the implementation of a binary search for the value of the input voltage. The charge-redistribution SAR works as follows:

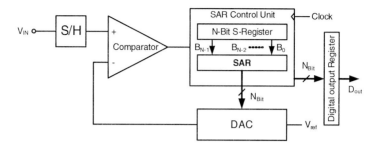

Figure 2.21: A block diagram of the SAR architecture.

1. Close the **Reset** switch to discharge the capacitor array and hence to set-up the offset voltage of the comparator ($V_{os} = V_{gnda}$) on that array.

2. Switch the input voltage, v_{IN}, onto the bottom plate of the capacitor array. The **Reset** switch is still closed.

3. Connect the bottom plate of the capacitor array to ground and open the **Reset** switch. As a result the voltage of the top plate of the capacitor array is then: $V_{Ctop} = V_{os} - v_{IN}$.

4. Switch the bottom plate of the MSB capacitor to V_{ref}. The output of the comparator represents the MSB of the digital value of the input voltage, v_{IN}. It is calculated as: $V_{Ctop} = V_{os} - v_{IN} + d_{N-1}\frac{V_{ref}}{2}$. If the comparator's output is high then the MSB capacitor is kept connected to V_{ref}. Otherwise, it is connected to ground.

5. Repeat the last step for the rest of the capacitors and modify the digital values in the SAR according to the output of the comparator.

The last conversion step for the last capacitor provides the following voltage on the top plate of the capacitor array: $V_{Ctop} = V_{os} - v_{IN} + d_{N-1}\frac{V_{ref}}{2} + d_{N-2}\frac{V_{ref}}{2} + ... + d_1\frac{V_{ref}}{2} + d_0\frac{V_{ref}}{2} \approx V_{os}$. The capacitor mismatch is one of the accuracy issues of this type of converters. This would affect mainly the INL and the DNL values of the ADC. Timing of this architecture is also sensitive as the input voltage faces a low pass filter generated by the switch resistance, R_{switch}, and the capacitor array, C_{sample}. The time constant of this filter is: $\tau = R_{switch}C_{sample}$. The voltage on the capacitor

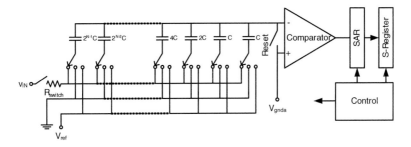

Figure 2.22: SAR converter using a binary weighted capacitor array.

array is then calculated by: $V_{sample} = v_{IN}(1 - \exp^{t/\tau})$. The previous formula can be used to determine the percentage of voltage on the capacitor array after a certain time. This is an important specification factor for the circuitry providing the input voltage, v_{IN}, to the ADC. The driver capability of this circuitry is essential for driving the capacitor array of the ADC within a certain time frame.

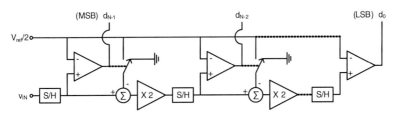

Figure 2.23: Block diagram of the pipeline ADC.

Pipeline Converters

The pipeline converters possess a relatively high resolution (up till 14 bits) and operate at high speed reaching frequencies in the 10^9 Hz range. It consists basically of N identical stages (one bit per stage) whereas each stage contains a comparator, a sample-and-hold, a summation component, and an amplifier with the gain of two (Fig. 2.23). The basic operation of the pipeline ADC is explained as follows:

- The input signal is compared to $V_{ref}/2$. The output depicts one bit conversion.

- The result of the conversion affects the input of the amplifier (gain = 2) as follows: if $v_{IN} > V_{ref}/2$ then V_{ref} is subtracted from the input of the stage. Otherwise the input is passed over to the amplifier.

- The value at the input of the amplifier is amplified by two and passed to the next stage.

The pipeline ADC needs N cycles to convert the first value. Afterwards one conversion is finished per clock cycle which results in a high throughput. The first stage of pipeline converters is crucial for the accuracy of the whole ADC as the error at this stage propagates throughout all the next stages.

Flash Converters

Flash converters are popular because of their high speed operation (one clock cycle per one digital output) though this high speed is at the cost of area and power consumption. Flash ADCs utilizes $2^N - 1$ comparators and 2^N resistors for N-bit resolution. Fig. 2.24 illustrate the architecture of a flash ADC. The reference voltage, V_{ref}, is divided into 2^N values (quantization levels) connected to the comparators which compare the quantization levels with the input voltage, v_{IN}. The N-bit digital output of the ADC is then determined by using a $2^N - 1 : N$ digital thermometer decoder. Mismatch of the resistor array restricts accuracy of flash ADCs. Therefore low resolution is usually used to keep the value of the INL less than 1/2 LSB.

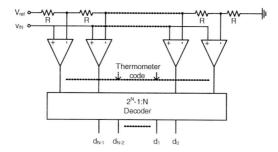

Figure 2.24: Block diagram of a flash ADC.

2.2.2.4 Wireless Power Transmission

There are mainly two methods for powering implanted systems: battery-powered and wireless-powered system. Battery-powered systems have a higher power efficiency as the losses are kept to a minimum compared to a wireless system with a large distance between a small implanted coil and a big external coil. However every battery has a certain life time. An exchange of the battery means a surgical intervention must take place. Surgical operations conclude risks for patients and possess a huge overhead (financially and operationally). Furthermore, surgical complications are more probable to occur during multiple surgeries which might lead to system failures due to improper implant dislocation, nerve injury, bleeding, mechanical complications, and others. A wireless-powered system offers a huge advantage by saving battery volume and reducing surgical intervention at the cost of losses in energy transmission.

2.2.2.5 Wireless Data Transmission

Implantable devices transmit data to an external receiver using a wireless link. There are several ways to implement the data transmission interface in an implantable system. Depending on the data rate and energy consumption of the implanted system it is possible to combine power and data transmission together. Digital modulation techniques are used to impress the digital signal on a carrier signal for transmitting data. High data rate requires high frequency of the carrier wave. With high frequency it is possible to transmit more energy though restricting factors are increased. For example, matching the sender and the receiver becomes more difficult for high frequency as the parasitic effects have more influence on the system, regulation of medical implant powering gets harder the higher the frequency is. Using low frequency restricts the amount of energy transmitted to the implantable system though it increases the flexibility of the system against imperfections in component matching. Separating wireless data and energy transmission frequencies leads to more flexibility in choosing the best frequency for the desired application according to existing conditions.

2.2.3   Signal Processing and Representation

Bio-signal processing is needed to indicate activity of neural or muscle tissue and hence to discover fatigue. Furthermore, a muscle signal processing is used to control a prosthesis. The algorithms for processing of bio-signals depend usually on the characteristics of the recorded signals, the recording method, and the placement of the electrodes. Invasive recordings conclude high selectivity as the sensors are placed close to the source of activity and possess a small recording area. Therefore motor unit conduction velocity is usually measured using high selective intramuscular electrodes. Surface recordings show high interference from neighboring tissues though separating the recorded wave into the main sources of potential (MUAPs) is rather difficult and time consuming. More details about signal processing are presented later in this work.

2.3   Packaging

Packaging of implantable devices is used to protect the body from toxic materials in those devices as well as to protect electronic implantable devices from body fluids. A hermetically-sealed and bio-compatible package is therefore needed to guarantee good isolation. Several materials can be used to package implantable devices. It is mainly dependent on the implantation period and the body tolerance to the packaging material.

Implantable electrodes are usually made out of bio-compatible metals such as gold, platinum, iridium, stainless steel. The electrode contacts and the wiring can be fabricated in a bio-compatible substrate such as polyimide or silicon.

Packaging of electronic systems is done by using bio-compatible materials to hermetically seal the system such as titanium, ceramics, glass, and silicon. These materials differ in their long term functionality. Silicon is used for short implantation period because body fluids reach the implant's electronics after weeks or maybe months of implantation. Metals, ceramics and glass offer a long-term characteristics and possess good bio-compatibility. Despite the good characteristics of metal encapsulations (hermetic packaging, bio-compatibility) they throw challenges for the efficiency of wireless power and data transmission.

2.4 Upper Extremity Prosthetics

Losing an upper extremity is a crucial incidence causing a huge restriction for casualties in daily life. The loss of upper extremity happens due to accidents, war or to amputation caused by trauma, disease, tumor or it is congenital [9]. In most cases patients use a state of the art artificial hand prosthetic. The development of upper extremity prosthetics has a long history, just a small part of it is known to us. The functionality level of the prosthetics is highly dependent on the available instruments and techniques during development time and on the needs of casualties. The 20th and 21st centuries depict a remarkable move in the quality of prosthetics due to modern versatile technologies and the manifold needs of casualties. Ambitious visions of researchers represent the key to success and development of our time.

Early prostheses were made out of metal because the patients themselves were mostly knights and warriors who lost an upper extremity. Metal limbs offered a restricted functionality of holding a sword or to break in a horse. The history of upper extremity prosthetics and its major development stations can be summarized as follows [54]:

- The Roman General Marcus Sergius had an iron hand prosthetic after he lost his right hand during the second Punic War (218-201 B.C.) [55].

- A mummy from the Ptolemaic period (305-30 B.C.) had an artificial arm after forearm amputation in order to support the victim with a new limb in the afterlife [56].

- There are several artificial limbs made of iron or steel from the 15th and 16th centuries [57]. The functionality of these hands was pretty much limited to holding reins whilst horse-riding or to hold a sword. The thumb was fixed in position whereas it was possible to move the fingers at once and fix them in a certain holding position. The prosthetic hand of Goetz von Berlichingen (a German knight from 1509) depicts an outstanding example of the hand prosthetics of the 16th century. Locking the fingers of the hand made it possible to hold a sword with better grip, as claimed, than normal hand [58].

- Bailiff, a German dental surgeon from Berlin of 19th century (1818), revolutionized the construction of prosthetics by involving the musculature of trunk in controlling the hand prosthetic. He used strings to keep the prosthetic closed. In order to open the prosthetic he used the trunk muscles [59].

- World War I and World War II increased the need to develop prostheses to supply huge number of war veterans. There were several forms of prostheses to meet the needs of injured soldiers who lost upper extremities. For the first time the outer shape (cosmetics to the outer surface of the prostheses) and the weight (plastic, carbon fibre) of the prostheses began to be important.

- During the second part of the 20th and the beginning of the 21st century the development of hand prostheses have been revolutionized. State of the art hand prostheses make use of bio-signals (muscle or nerve signals) or a body part (the leg or one healthy hand) to control a mechanical/electrical hand prosthesis. Furthermore, a sensory feedback method into the nervous system have been frequently used to provide a tactile sense from the arm. Bio-signal analysis has strongly focused on the development of prosthetic systems with several degrees of freedom. As a result, the need of various independent control signals made it necessary to develop bio-potential recording systems of high quality and wide range specification.

Most state of the art prostheses use the contraction of muscles to open and close a prosthetic hand. The muscles act as an amplifier to the actual control signals coming from the nerves. Several commercial prosthetic systems controlled the grip force of a hand prosthetic by the intensity of the muscle signal like the early Otto Bock hand prosthetic [60]. Hudgin et al. [61] suggested a new strategy towards feature based controlling of a multi-functional myo-electric prosthesis. The strategy contains mainly methods for time domain features extracted from the EMG signal. A more precise analysis of EMG signals utilizes time and frequency representation of the muscle signals which in turn delivers more information about the desired physical activity. Some prosthesis designs utilize the data directly from the brain to control robot arms [62]. Experiments on rhesus monkeys showed that motion of the arm can be predicted a good tenth of a second prior to muscle movement with recorded signals from less than a 100 brain cells of the primary motor cortex [63] though signal acquisition from deep brain cells and huge processing overhead are the main obstacles for practical use of such methods.

Despite the advances in actual hand prostheses the development of an intuitive alternative to the natural hand is not available yet. The nerves are the natural controller of our muscles. Therefore, feeding a hand prosthetic directly with nerve signals from the brain would be the most natural way for intuitive control of future hand prostheses. However, there is still a long way to go until such a hand prosthesis can be put on the market. Sensors with high precision, small area, and high resolution need to be developed in order to collect selectively the nerve signals in a dense structure (a bundle of 70 thousand nerve cells controls the hand muscles). Furthermore, the processing of a big amount of data from neural sensors and separating these signals into their independent components to extract different movement structures under real time consideration and low power operation would be essential for intuitive prosthesis.

Nowadays hand prostheses have a modular structure (future update with additional components is possible) with optimal safety and reliability for user. They even create a natural appearance to the human skin. Otto Bock's [60] bionic hand Michelangelo weighs about 420 gr and possesses several gripping force modes (15N, 60N and 70N). Moreover, it contains a control mechanism with multiple degrees of freedom (multi-axial movement with about 6 DoFs) based on myo-electric data. However, a sEMG recording is restricted in degrees of freedom to control a hand prosthetic system smoothly and reliably. The advances in EMG signal processing methods and in nerve/muscle surgery (TMR: trans-muscular re-innervation) made it again possible to make some improvement in development of hand prosthetics. The problem of an appropriate feedback to the body is still a matter of research and needs to be further developed in order to regain part of the natural functionality of the hand. There are other prosthetic hand products nowadays with several setting possibilities like the i-limb ultra hand prosthetic of Touch Bionics [64]. The i-limb has multiple modes that can be automatically selected and trained using a software module (biosim-i or biosim-pro) which shows the lack of intuitive control of the movement. BeBionic Hand [65] also offers a myo-electric hand prosthetic with several grip patterns optimized by wireless programming procedure. The movement of the prosthetic in a certain grip mode looks natural. But the switch to a different mode needs to be done by the user. The switching between different modes increases the capabilities of present-day hands. But the intuitive transition between these modes was not accomplished yet. To controlling the mechanics smoothly and intuitively more information and more independent signals from the bio-recorded data are needed. This task has two main requirements:

1. Improvements in the field of bio-sensors and acquisition methods, and

2. improvements in decoding and processing of recorded bio-data

The draw-backs of present-day hand prosthetic systems can be summarized as
follows:

- Fewer degrees of freedom compared to robotic hands.

- The lack of an intuitive control mechanism which depicts the need of training
  and concentration whilst using the prosthetic. That limits the concentration
  on other things than the prosthetic itself. A robust complex intuitive control
  of the prosthesis would revolutionize the production of hand prostheses.

- Battery life (less essential).

- Grasp and movement of the hand is limited to some modes and options activated by the user. People who lost both their hands will face a problem of
  switching the several modes present in the prosthesis.

- Time delay: myo-electric prostheses show a time delay during function. (The
  porting of some EMG data analysis to the acquisition system might improve
  the real time condition of the prosthetic system...)

- Poor feedback to the body: Feedback to the body must be part of future prostheses. A proper feedback is essential regarding safety aspects and adjustable
  control of the prosthetic.

- Cost: despite the low functionality level present day prosthetics are expensive.

- Low area resolution of the recorded EMG, which results in low number of DoFs.
  Surface electrodes possess a big area in order to reduce impedance. Using big
  electrodes might increase the correlation factor between several recording electrodes on the skin surface. This reduces the selectivity of the EMG signals
  recorded by these electrodes and therefore makes it difficult to regain independent signal components with easy algorithms.

- The surface recording of the EMG signals causes irritations on skin and it lacks the long term stability due to several artifacts (movement artifacts, electromagnetic coupling, electrode artifacts, non-consistent electrode-skin interface, etc.).

Chapter 3

Implant System

The MyoPlant[1] project provides an implantable system approach towards improving the functionality of a hand prosthetic system. The utilization of implantable solution delivers a better view of the recorded signals, as the electrodes are closer to the source signals, and might improve the functionality of the hand prosthesis by delivering good shaped control signals consistently.

The advantages of using an implanted system can be summarized as follows:

- Better EMG resolution according to the placement advantages directly on the muscle tissue.

- High selectivity through retrieval of independent signals from neighboring muscles.

- Avoid skin irritations due to surface electrodes.

- Better signal quality (better SNR). The analogue signal path is reduced and hence fewer noise sources are collected through wires. Besides, higher signal amplitude can be recorded.

- Improvement of signal stability over time for long-term recordings due to stable electrode-tissue interface.

- Insight into high frequency portions of the EMG signals which are filtered using surface electrodes due to low-pass filter behavior of the skin.

---

[1]Development of a Bionic Hand Prosthesis on the Basis of a **MYO**genic Intelligent Im**PLANT** system (BMBF: 16SV3699).

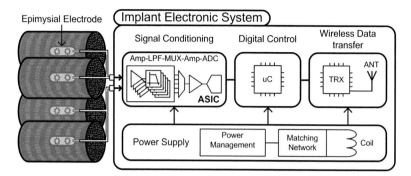

Figure 3.1: Block diagram of the implant system.

The MYOPLANT system consists basically of two main parts:

- Implant system: The implant system is used to record the EMG signals by means of implantable electrodes, to amplify the low-voltage muscle signals, to digitize them, and to transmit these signals via wireless link to the external part of the system.

- External system: The external system consists mainly of the receiver interface to communicate with the implanted device and an external coil to power the implanted system inductively.

The implant system consists of two main parts: implantable electrodes, and an implant electronic system encapsulated with a biocompatible material. Fig. 3.1 illustrates a block diagram of the designed system. Bipolar silicon electrodes are placed sub-epimysially under the epimysium layer of the muscle for collecting EMG activity as closely as possible to the signal origin. The electronic system concludes interfaces for wireless energy and data transmission. The implantable electronic system is used to capture EMG signals from the epimysial electrodes, amplify the low-voltage muscle signals, digitize them, and finally to transmit these signals via wireless link to an external module for further signal processing.

3.1   Implantable Epimysial Electrodes

The main problems of implantable electrodes are a high electrode impedance, which reduces signal to noise ratio (SNR) and hence the resolution of the input signal, and the mechanical stability in a continuously contracting and relaxing muscle. Especially the contacting position of the wire to the electrode is usually very sensitive to mechanical stress. Several types of implantable electrodes (epimysial electrodes) were probed for the invasive recording of muscle activity. A flexible electrode structure using polyimide substrate and platinum contacts [5] was tested in an animal experiment. The electrode impedance was reduced due to the rough structure of the surface. An evaluation study showed suitability for EMG recording [66] though the bonding of the wire to the electrode contact is vulnerable to strong muscle movement. Silicon epimysial electrodes with higher volume showed better mechanical stability during implantation. The epimysial silicon electrodes and the implantation procedure were designed and successfully evaluated in a rat and a sheep model by Lewis et al. [67]. The carrier of the silicon electrode has a thickness of 0.9 - 1.1 mm. The electrode contact disks are made of stainless steel with a surface area of 7.1 mm² and separated by 1 cm (Fig. 3.2). The electrodes were sub-epimysially tunneled and then fixed through sutures on the electrode carrier. Following muscles were targeted in an implantation in a sheep to validate functionality of the implant system: (a) musculus brachiocephalicus (b) musculus latissimus dorsi (c) musculus triceps brachii (d) and musculus brachialis.

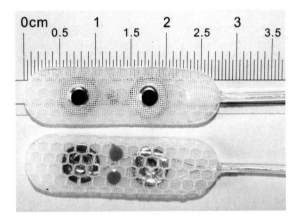

Figure 3.2: Epimysial silicon electrode: back view (top), and front view (bottom). The silicon epimysial electrodes were fabricated at the Fraunhofer Institute for Biomedical Engineering (IBMT).

The electrode interface and especially its impedance usually affects the SNR of the input signal to the conditioning circuit. The resolution of the electronic system cannot exceed that of the input signal. This piece of information is most important for specifying several parameters of the acquisition electronic system such as resolution and noise. The electrode noise has several sources such as the thermal noise by the real impedance of the electrode interface, microphonics noise caused by the motion of the electrode or the fluid underneath, shot noise induced by electron motion, and popcorn noise generated by modifications of the electrode surface.

Considering thermal noise at the electrode tissue interface (ETI), the mean-square noise voltage is calculated by the Nyquist equation [37] as follows:

$$\overline{v_{th,el}^2} = 4kTR \cdot \Delta f \tag{3.1}$$

where k is the Boltzmann constant ($1.38 \cdot 10^{-23} \text{JK}^{-1}$), T is the temperature (in Kelvin), R is the resistance (in ohms), and Δf is the frequency bandwidth. Noise magnitude at the electrode is inversely proportional to the square root of the area of the electrode (noise $\propto \frac{1}{\sqrt{A}}$) [39]. Therefore, increasing the effective area increases the SNR of the recorded signals and hence the quality of the EMG recording. The SNR of the input signal is needed for the design specification of the electronic circuit (resolution of the ADC. See section 2.2.2.3, equation 2.16).

A simple equivalent circuit of the electrode-tissue interface (ETI) [68] is often adequate for an approximation of the thermal noise level at the ETI. Fig. 3.3 illustrates the equivalent circuit of the ETI and its equivalent noise model. The impedance of the equivalent circuit is given as:

$$Z_{R_F \| C_H} = \frac{R_F}{1 + j\omega C_H R_F} = \frac{R_F(1 - j\omega R_F C_H)}{1 + \omega^2 C_H^2 R_F^2} = \frac{R_F}{1 + \omega^2 C_H^2 R_F^2} + j\frac{-\omega R_F^2 C_H}{1 + \omega^2 C_H^2 R_F^2} \tag{3.2}$$

$$Z_{ETI} = Z_{R_F C_H} + R_E = \left[R_E + \frac{R_F}{1 + \omega^2 C_H^2 R_F^2} \right] + j \left[\frac{-\omega R_F^2 C_H}{1 + \omega^2 C_H^2 R_F^2} \right] \tag{3.3}$$

The referring thermal noise voltage per square-root Hertz results in:

$$\frac{v_n}{\sqrt{\Delta f}} = \sqrt{4 \cdot k \cdot T \cdot Re(Z_{ETI})} \tag{3.4}$$

where v_n is the rms noise voltage, Δf is the signal bandwidth, k is the Boltzmann constant, T is the temperature (Kelvin), and $Re(Z)$ is the real part of the impedance of the ETI. Using in vivo measured values after 12 weeks from the sheep implantation [69], the rms thermal noise voltage of every single contact of the silicon electrode is

$v_n = 0.83 \ \mu V_{rms}$ over a bandwidth of $\Delta f = 7...1500$ Hz. The values of the equivalent circuit components in Fig. 3.3 are: $R_E = 4 \ k\Omega$, $R_F = 70 \ k\Omega$, and $C_H = 1.2 \ \mu F$.

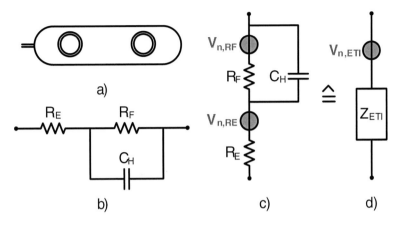

Figure 3.3: Equivalent circuit of the implantable epimysial electrode. a) Epimysial electrode. b) Simple equivalent circuit of the electrode-tissue interface (ETI). c) Noise model of the ETI. d) Simplified noise model of the ETI.

3.2 Implant Electronic System

The implantable electronic system is used to differentially acquire EMG signals collected by the epimysially implantable electrodes, amplify the low-voltage muscle signals, digitize them, and to transmit these signals per wireless link to an external part of the system. A wireless power supply per inductive coupling at 125 kHz is used to provide the implant electrical system with the needed power for operation. Several voltages are needed for the several chips of the implant system. Therefore two main voltage regulators and one voltage divider are used for voltage management. The EMG signals in digital form are sent via a transceiver for wireless data transmission utilizing the MICS band (402 MHz) for biomedical applications [70]. A microcontroller is used to control the communication and initialization between the chips on the implant system and with the external application. Fig. 3.4 illustrates the implanted module in silicon package. The implant module energy and data antenna was designed and fabricated by the Fraunhofer Institute for Biomedical

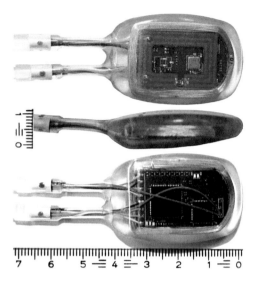

Figure 3.4: Implant electronic system in a silicone package. Top, bottom, and collateral view starting from the top respectively. The implant electronics were capsuled in silicone at the Center for Medical Physics and Biomedical Engineering, Medical University of Vienna.

Engineering (IBMT), Cardona et al. [71].

   Detailed description of the sub-blocks of the implant system is presented in the upcoming chapters.

Chapter 4

Signal Conditioning Circuit

A mixed-signal chip in 130nm technology is described in this chapter. The chip is designed to record invasive muscle activity in order to deliver an input for the controller of a hand prosthetic. It acquires EMG signals from five differential inputs. The signals are amplified, multiplexed and digitized inside the developed chip. The ASIC is optimised for low power and low noise and is intended for application in medical implants. Furthermore, the chip provides configuration options for several parameters for the analogue part of the chip (gain setting, channel power on/off etc.). The chip has been tested in an implantable system for recording EMG signals from implanted epimysial electrodes. The digitized data samples are collected by a microcontroller on the implant system, which sends them to a telemetry chip for wireless transmission to the external system of the hand prosthesis.

Volume, power consumption, low noise and efficiency play an essential role during the development of an implantable system. A trade-off between these parameters is often inevitable for the developer. A transistor technology with small feature size has many advantages considering area reduction and power consumption. At the same time, the same technology may have strong drawbacks in terms of noise contribution and amplification factor.

4.1 System Requirements and Specification

Muscle cells have a potential difference of $\approx$ 100 mV between the intra- and extracellular medium [72], while implanted electrodes record amplitudes in the range of 100 μV - 5 mV. Therefore, a system for amplifying these small bio-signals is needed. The impedance of the used electrodes and the strength of the bio-signal are important values for the dimensioning of the bio-amplifier. The electrode interface and the bio-amplifier input constitute two impedances in series connected to the bio source signal (Fig. 2.12). This implies a voltage divider for the signal source. The input impedance of the bio-amplifier should be much larger than the impedance of the electrode interface in order to obtain a large voltage drop on the input of the amplifier. The bio-amplifier also needs to accommodate the voltage offsets of the electrode-tissue interface. In order to overcome the problem of a comparatively large DC-offset on the differential input of the amplifier, which might drive the amplifier into saturation, an AC-coupling method was used for this stage [49]. The ASIC has an analogue to digital converter (ADC) for digitizing the recorded signals. The resolution of the ADC is set as a trade-off between the target signal-to-noise ratio and the allowable data rate by the telemetry chip. The use of an analogue filter reduces the data rate while using a high resolution at the ADC. Furthermore, the analogue filter removes irrelevant signal frequencies by limiting the bandwidth to the needed EMG-bandwidth for controlling the prosthesis. A configurable sampling rate at the ADC allows us to change the data rate sent by the telemetry chip according to link quality. A well defined specification is very essential before starting the development of an ASIC. The main requirements of the developed ASIC are therefore summarized as follows:

- Low power consumption;

- Low noise;

- High common mode rejection ratio (CMRR);

- Small silicon area;

- Configurable data rate to accommodate the quality of the wireless link;

- Configurable gain setting for different source signal amplitudes;

- DC-cancellation;

Several methods for acquisition of bio-signals (unipolar, bipolar or tripolar recordings) are reported in literature. The chip was constructed for bipolar acquisition method (section 2.2.1). The basic advantages of this method are as follows:

- Better symmetry in the input circuit;

- No single point of failure (shared reference in the case of unipolar recording);

- Surgically easier to perform;

- Less vulnerable to interference from neighboring or from deep muscles (better rejects common mode noise).

  On the other hand, the bipolar acquisition method needs more electrodes and contacting wires than the unipolar method.

An initial specification of the chip was set prior to design work. Several parameters were set according to the requirements of the prosthetic control application, such as signal bandwidth and resolution values. The rest was set according to literature research and circuit conditions. Table 4.1 presents the specification parameters for the ASIC.

Table 4.1: ASIC Specification.

Parameter	Value	Unit	Comment
Number of channels	5		
Input range	± 0.5 ± 1 ± 5 ± 12	mV	Externally configurable
Bandwidth	$100 - 800$ $100 - 1500$	Hz	Externally configurable
Input-referred noise	3 4	μV_{rms}	$100 - 800$ Hz $100 - 1500$ Hz
CMRR	50	dB	@ 50 Hz
Resolution	10	Bit	

4.2 Architecture and Design

A block diagram of the ASIC is shown in Fig. 4.1. It has five analogue channels, each of which consists of a preamplifier and a low-pass filter followed by a driver. As

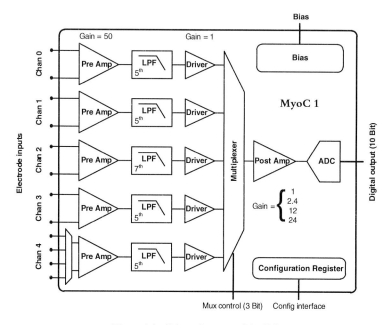

Figure 4.1: Chip architecture (MyoC1).

a concept study, the MyoC1-chip implements different circuit alternatives in each channel. Channel 4 has a multiplexer at the input in order to test the functionality of switching between two electrode pair to increase system reliability by connecting multiple electrode pairs per recording channel. Channel 2 utilizes a 7th order low-pass filter for testing the benefit of a higher filter order. The channels are combined by another multiplexer, which directs the selected signal to a programmable post-amplifier. An analogue to digital converter (ADC) following the amplifier provides the digitized signal at a 10-bit parallel port. A configuration register with a serial interface is used for setting the configuration parameters of the chip. Further control lines are used to select the active channel on the multiplexer (Mux control).

The multiplexed architecture provides advantages for lowering power consumption by using single ADC for a certain number of channels (five channels in the MyoC1-chip) and a simple low power components at the analogue front end. The post-amplifier provides further amplification for better resolution and hence better signal to noise ratio at the output.

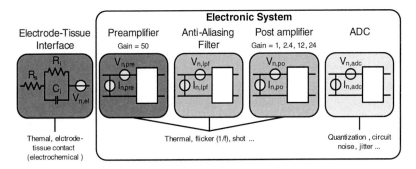

Figure 4.2: Noise sources in the system for EMG signal acquisition.

4.2.1 System Noise Analysis

Electronic noise limits the precision and hence the quality of a biopotential recorded signal. A macro noise analysis is needed to quantify the permitted noise for every single component in the measurement system and specify the needed signal-to-noise ratio at the output of the system. The SNR is an important quantity which directly affects the power consumption and the area of a conditioning circuit for medical application as seen later in this chapter. Low noise design mostly acquires large capacitors and large transconductances (g_m), which increases power consumption. Therefore an accurate specification of the signal-to-noise ratio and hence the permitted electronic noise is essential for an electronic circuit designer. Noise in a biopotential recording system has multiple origins. An in-system noise emerges in the system components (Fig. 4.2) such as electrodes and conditioning circuits (thermal and flicker noise). This noise is modeled as a stationary process (no time change) and have a zero mean (zero dc content). Circuit design strategies are used to optimize in-system noise to acceptable levels. A simplified analysis of the noise sources depicted in Fig. 4.2 is given as:

$$v_{ni,total}^2 = v_{ni,el}^2 + v_{ni,pre}^2 + \frac{v_{ni,lpf}^2}{A_{pre}^2} + \frac{v_{ni,po}^2}{A_{pre}^2} + \frac{v_{ni,adc}^2}{A_{pre}^2 \cdot A_{post}^2} \qquad (4.1)$$

where $v_{ni,x}^2$ represents the input-referred noise voltage power, and A_x is the gain of component x. Equation 4.1 reveals the noise sources in the designed system and their dependencies. Electrode interface and the preamplifier (in red) are the most critical components considering noise in the acquisition system. The noise

contribution from the anti-aliasing filter, from the post-amplifier and from the ADC
is reduced due to the amplification factor of the preamplifier. The electronic system
noise[1]

$$\overline{v_{ni,ES}^2} = \overline{v_{ni,pre}^2} + \frac{\overline{v_{ni,lpf}^2}}{A_{pre}^2} + \frac{\overline{v_{ni,po}^2}}{A_{pre}^2} + \frac{\overline{v_{ni,adc}^2}}{A_{pre}^2 \cdot A_{post}^2} \tag{4.2}$$

can be neglected from equation 4.1 when $\overline{v_{ni,ES}^2} \ll \overline{v_{ni,el,2}^2}$ is accomplished. For
example if $V_{ni,ES} = V_{ni,el}/2$ then the noise power from the electronic system builds
25% of the total noise power from the electrode tissue interface.

Coherent interference (CI) is another source for noise in bio-potential acquisition systems. It has external origins such as line noise (50/60 Hz), electromagnetic
interference (EMI) from electrical devices in the surrounding or sender antennas.
Shielding and filtering techniques can be used to reduce CI effects in the medical
recording system. Besides, fully differential input circuits and certain layout techniques are used to reduce common noise at the inputs of an amplifier circuit.

The signal at the input of the electronic system already concludes a noise content
mainly from the electrode-tissue interface. The electronic system must keep its
noise level under the input noise value generated by the electrode-tissue interface to
reduce the overall noise in the system (section 2.2.2.1). The electrode noise from one
contact at the ETI has been calculated in equation 3.4. In a differential biopotential
recording two such sources must be considered. Fig. 4.3 depicts the noise sources of
the electrode-tissue interface and those of an input amplifier in a recording system.

As noise voltages add in a root sum of squares manner (assuming they are uncorrelated with each other. See section 2.2.2.1), the noise contribution from the two
electrode contacts is calculated as follows:

$$v_{sn,in} = \sqrt{v_{sn1}^2 + v_{sn2}^2} = \sqrt{2} v_{sn,el} \tag{4.3}$$

where $v_{sn,el}$ is the thermal noise of the electrode-tissue interface according to equation 3.4.

The noise behavior at the ETI cannot be described only by the thermal noise.
There are other noise sources accumulating to the thermal noise at the electrode-
tissue interface such as **microphonics** (caused by the motion of the electrode or the

---

[1]Electronic system noise: The input-referred noise from the preamplifier, anti-aliasing filter,
post-amplifier, and from the ADC.

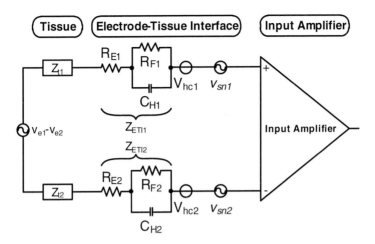

Figure 4.3: Schematic diagram of the input amplifier including noise sources from the electrode-tissue interface (ETI). $v_{e1,2}$ is the differential input signal from two electrodes, $Z_{t1,2}$ is the tissue impedance (mostly resistive component), $Z_{ETI1,2}$ is the equivalent impedance of the ETI, $V_{hc1,2}$ is the half-cell potential (DC voltage), and $v_{sn1,2}$ is the thermal noise of the ETI.

fluid underneath), **shot noise** (induced by electron motion), and **popcorn noise** (generated by modifications of the electrode surface). Fernandez et al. [73] reported a root-mean-square noise voltages from 1 μV_{rms} to 15 μV_{rms} at the body surface using Ag-AgCl electrodes.

The noise for the analogue channel in the MyoC1 ASIC was specified to be 3 μV_{rms} and 4 μV_{rms} on a bandwidth of 100 - 800 Hz and 100 - 1600 Hz respectively. The noise was specified to be smaller than the electrochemical noise of the used electrodes and therefore does not affect the signal quality as most of the noise already comes from the electrode interface. Furthermore it represents a trade-off between signal-to-noise ratio and power consumption and area in the ASIC because reducing the noise costs more power and more area which must be kept to a minimum in implantable devices.

4.2.2   Analogue Front End

The analogue front-end consists of the preamplifier, low-pass filter, multiplexer, and post-amplifier. EMG amplifier gains depend on the EMG input range recorded internally (Table 2.1). A configurable gain up to 60 dB is usually needed for good resolution of the input signal. The bandwidths of EMG amplifiers depict the transient nature of the recorded MUAPs. An adjustable recording bandwidth provides an opportunity for better signal-to-noise ratio by gradually decreasing the recorded signal to the band of interest and hence rejecting outer band components. This is usually dependent on the external application which determines the characteristics of the signal it processes. The design of each element of the analogue front-end is described in the following sections.

4.2.2.1   Preamplifier

Fig. 4.4 depicts a circuit diagram of the operational transconductance amplifier (OTA) used in the preamplifier (appendix B.2 contains the transistor sizing of the designed OTA). The OTA has low impedance at all nodes except the input and output nodes which results in low driver capability suitable for capacitive loads. The symmetry of the OTA improves the matching and hence the CMRR and offset characteristics of the OTA.

The specification of the amplifier is summarized in Table 4.2.

Table 4.2: Specification of the preamplifier.

Parameter	Value
V_{dd}	1.2 V
I_{Bias}	$< 2~\mu A$
Temperature	$37^\circ C$
Closed loop gain	34 dB
GBW	> 100 kHz
CMRR	> 60 dB
Output voltage swing	≈ 600 mV
Noise	$< 2~\mu V_{rms}$
Phase margin	$> 45^\circ$

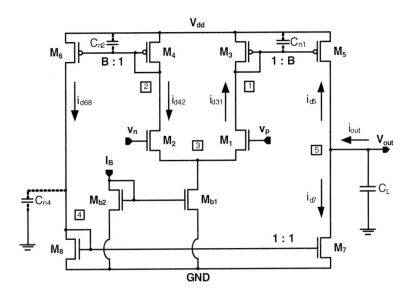

Figure 4.4: The symmetrical OTA used in the preamplifier.

Operating Region

Scaling down MOS transistor dimensions reduces the supply voltage and the saturation voltage, V_{dsat}, of the transistors. This drives the transistors excessively towards moderate inversion and weak inversion regions ($V_{GS} < V_{th}$) for low power operation. The preamplifier's OTA must work in the micropower range to fulfill the specification of low power for implantable devices. An OTA operating in the weak inversion uses low-voltage and low current and hence consumes low power. The drain current in weak inversion has an exponential form instead of the square-law form of the current in strong inversion. It is mostly a diffusion current in which electrons diffuse from higher carrier concentration (source region) to lower carrier concentration (drain region). More information about the behavior of MOS transistors in weak inversion is to be found in B.1.

The transconductance g_m of a MOS transistor operating in the weak inversion region is proportional to the drain current as derived in 4.5. This is similar to bipolar transistor behavior[2].

$$I_D = \frac{W}{L} I_{D0} exp(\frac{V_{GS}}{nkT/q}) \tag{4.4}$$

$$g_m = \frac{dI_D}{dV_{GS}} = \frac{I_D}{nkT/q} \tag{4.5}$$

where I_{D0} is process dependent, n is the subthreshold factor (technology dependent: $1.2 \cdots 1.4$), k is the Boltzmann constant ($1.38 \cdot 10^{-23}$ Ws/K), T is the temperature in Kelvin (body temperature is $37°C = 310.15$ K), and q is the charge of an electron ($1.602 \cdot 10^{-19}$ C). As a result, the bias current is used to directly control the transconductance of the MOSFET. Very low currents can be used to accomplish reasonable values for the transconductance g_m for a desired gain of the preamplifier or a GBW.

Operation

As mentioned before, only the input and output nodes are of high-impedance. The input devices and the load transistors are sized equally ($W_1/L_1 = W_2/L_2$ and $W_3/L_3 = W_4/L_4$). The differential input results in two differential currents, i_{d31} and i_{d42}. They are calculated in 4.6.

$$-i_{d31} = i_{d42} = \frac{g_{m1}}{2}(v_p - v_n) = i_d \tag{4.6}$$

According to the sizing factor B the mirrored current into M_5 and into M_6 equals to $i_{d5} = -i_{d7} = B \cdot i_{d31} = -B \cdot i_{d42}$. As a result, the output voltage is calculated as:

$$v_{out} = 2Bi_d(r_{o5}||r_{o7}) \tag{4.7}$$

The (open loop) voltage gain of the OTA is then calculated as:

$$A_v = \frac{v_{out}}{v_p - v_n} = B \cdot g_m \cdot (r_{o5}||r_{o7}) \tag{4.8}$$

The output current is calculated as follows:

$$i_{out} = i_{d5} - i_{d7} = 2Bi_d \tag{4.9}$$

---

[2]The collector current of *npn* bipolar transistor is: $I_C = I_S \cdot exp(V_{BE}/U_T)$, where I_S is the current, V_{BE} is the base-to-emitter voltage, and U_T is the thermal voltage ($U_T = kT/q$). The transconductance of a bipolar transistor is: $g_m = I_C/U_T$.

The transconductance of the OTA is then calculated as:

$$g_{mOTA} = \frac{i_{out}}{v_p - v_n} = B \cdot g_m \tag{4.10}$$

The unity-gain frequency (gain = 0 dB) is given in 4.13.

$$v_{out} = i_{out} \cdot \frac{1}{j\omega C_L} = B \cdot g_m(v_p - v_n) \cdot \frac{1}{j\omega C_L} \tag{4.11}$$

$$\frac{v_{out}}{v_p - v_n} = \frac{B \cdot g_m}{j\omega C_L} \rightarrow \left| \frac{v_{out}}{v_p - v_n} \right| = \frac{B \cdot g_m}{2\pi f \cdot C_L} \tag{4.12}$$

$$\frac{B \cdot g_m}{2\pi f \cdot C_L} = 1 \rightarrow f_{un} = f_{0dB} = \frac{B \cdot g_m}{2\pi C_L} \tag{4.13}$$

There are mainly two important poles for the OTA. The first pole frequency f_{3dB} is in 4.14. It defines the bandwidth of the open loop gain. A second non-dominant pole is located at node 1 (Fig. 4.4).

$$f_{3dB} = \frac{1}{2\pi(r_{o5}||r_{o7})C_L} \tag{4.14}$$

$$f_{nd} = \frac{g_{m3}}{2\pi C_{n1}}, \quad C_{n1} = (1+B)C_{GS3} + C_{DB3} + C_{DB1} \approx (3+B)C_{GS3} \tag{4.15}$$

A pole-zero is located at node 4 (C_{n4} results in a pole and a zero). Therefore, the phase margin is not affected by this node.

Noise Analysis

Thermal noise is generated due to the channel resistance, R_{CH}, of the MOSFET as a result of random motion of electrons due to thermal effects. The channel resistance can be modeled as $1/R_{CH} = \gamma g_m$, where γ is a coefficient [74]. $\gamma = 2/5$ for long channel MOSFETs and it reaches $\gamma \cong 1$ for short channel devices [75]. The PSD of the output current of the thermal noise and its input-referred counterpart are defined as follows:

$$S_{Id,th} = 4kT\gamma g_m \tag{4.16}$$

$$S_{Vg,th} = 4kT\frac{\gamma}{g_m} \tag{4.17}$$

As a result we get for the power of the output noise current and of the input-referred noise voltage:

$$I_{on,thermal}^2 = S_{Id,th}\Delta f = 4kT\gamma g_m\Delta f \tag{4.18}$$

$$V_{in,thermal}^2 = S_{Vg,th}\Delta f = \frac{I_{on,thermal}^2}{g_m^2} = 4kT\frac{\gamma}{g_m}\Delta f \tag{4.19}$$

In the low frequency range of analogue devices flicker noise[3] is dominant. It is inversely dependent on the frequency and caused by the random trapping and detrapping of mobile carriers in the traps at the $Si - SiO_2$ interface and also at the gate oxide. Furthermore, fluctuations in the carrier mobility and defects in the crystalline structure during fabrication [76] contributes to flicker noise. Down-scaling of transistor size degrades the noise characteristics of transistors. Chew et. al. [77] reported an increase by 1.5 order in the noise spectral density S_{Id} of thin gate oxide transistors when scaling from 350 nm to 130 nm.

Flicker noise is inversely proportional to the size of a MOSFET as shown in the equations 4.22 and 4.23. The PSD of the flicker noise current/voltage can be modeled as follows [51]:

$$S_{Id,flicker} = \frac{KF \cdot I_d^{AF}}{f \cdot (C'_{ox})^2 LW} \tag{4.20}$$

$$S_{Vg,flicker} = \frac{KF \cdot I_d^{AF}}{f \cdot (C'_{ox})^2 LW \cdot g_m^2} \tag{4.21}$$

As a result we get for the power of the output noise current and of the input-referred noise voltage:

$$I_{on,flicker}^2(f) = S_{Id,flicker}\Delta f = \frac{KF \cdot I_d^{AF}}{f \cdot (C'_{ox})^2 LW}\Delta f \tag{4.22}$$

$$V_{in,flicker}^2(f) = S_{Vg,flicker}\Delta f = \frac{KF \cdot I_d^{AF}}{f \cdot (C'_{ox})^2 LW \cdot g_m^2}\Delta f \tag{4.23}$$

where KF is the flicker noise coefficient, I_d is the drain current, AF is the flicker noise exponent ($\approx 0.5 \cdots 2$), and f is the frequency.

The overall noise of the MOSFET low frequency operation can be modeled as follows:

$$V_{in}^2 = V_{in,thermal}^2 + V_{in,flicker}^2 = 4kT\frac{\gamma}{g_m}\Delta f + \frac{KF \cdot I_d^{AF}}{f \cdot (C'_{ox})^2 LW \cdot g_m^2}\Delta f \tag{4.24}$$

There are two parameters which can be used to reduce the noise in the MOSFET devices operating in the weak inversion:

- **Drain current:** the transconductance g_m correlates with the drain current. Increasing the drain current (bias current) leads to a $\sim 1/I_d^2$ reduction in the flicker noise and to $1/I_d$ reduction in the thermal noise power.

---

[3]Flicker noise is also called $1/f$ noise.

- **MOSFET gate size:** The area of the MOSFET gate correlates inversely to the flicker noise power in the MOSFET.

As stated before, the noise at the input of the system is maximally 15 μV_{rms} (electrode noise). The noise specification of the preamplifier was set to be $< 2\ \mu V_{rms}$ to avoid the unwanted increase of overall noise of the recording system. Deriving the noise formula for the symmetrical OTA in Fig. 4.4 is necessary for the designer. The noise contribution of every device in the OTA has to be divided by g_{m1} (the transconductance of the input devices) to be referred to the input. Fig. 4.5 illustrates the noise contribution at the input stage.

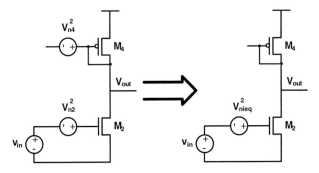

Figure 4.5: Noise of the input stage with active load.

$$I_{n,out}^2 = g_{m2}^2 \cdot V_{n2}^2 + g_{m4}^2 \cdot V_{n4}^2 \tag{4.25}$$

$$V_{nieq}^2 = \frac{I_{n,out}^2}{g_{m2}^2} = V_{n2}^2 + V_{n4}^2 \cdot \left(\frac{g_{m4}}{g_{m2}}\right)^2 \tag{4.26}$$

The output noise of a current mirror is depicted in Fig. 4.6. The power of the output noise current in the NMOS current mirror is calculated as follows:

$$I_{nout}^2 = I_{n7}^2 + B^2(I_{ni}^2 + I_{n8}^2) \tag{4.27}$$

A large current mirror factor would increase the output noise of the circuit. Therefore a factor of B = 1 was chosen.

The noise contribution of all devices of the OTA is given by:

$$V_{nieq}^2 = 2 \cdot V_{n1}^2 + 4 \cdot V_{n3}^2 \cdot \left(\frac{g_{m3}}{g_{m1}}\right)^2 + 2 \cdot V_{n7}^2 \cdot \left(\frac{g_{m7}}{g_{m1}}\right)^2$$

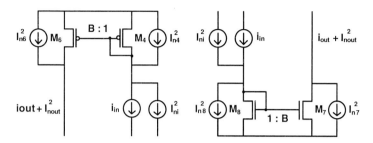

Figure 4.6: Noise of a current mirror.

$$V^2_{nieq} = 2 \cdot V^2_{n1}\left[1 + 2 \cdot \left(\frac{V^2_{n3}}{V^2_{n1}}\right)\left(\frac{g_{m3}}{g_{m1}}\right)^2 + \left(\frac{V^2_{n7}}{V^2_{n1}}\right)\left(\frac{g_{m7}}{g_{m1}}\right)^2\right] \qquad (4.28)$$

in which $M_3 - M_6$, $M_7 - M_8$, and $M_1 - M_2$ are equally sized respectively and the current mirror multiplier B = 1.

Equation 4.28 for thermal noise components is derived as:

$$V^2_{nieq,thermal} = \frac{8kT\gamma}{g_{m1}}\left[1 + 2 \cdot \left(\frac{g_{m3}}{g_{m1}}\right) + \left(\frac{g_{m7}}{g_{m1}}\right)\right]\Delta f \qquad (4.29)$$

The input-referred flicker noise is given by:

$$V^2_{nieq,flicker} = \frac{2 \cdot KF_1 \cdot I^{AF}_{d1}}{f \cdot (C'_{ox1})^2 L_1 W_1 \cdot g^2_{m1}}\left[1 + 2 \cdot \frac{KF_3 \cdot C'^2_{ox1} L_1 W_1}{KF_1 \cdot C'^2_{ox3} L_3 W_3} + \frac{KF_7 \cdot C'^2_{ox1} L_1 W_1}{KF_1 \cdot C'^2_{ox7} L_7 W_7}\right] \cdot \Delta f$$
$$(4.30)$$

The transconductance of the input devices must be greater than that of the rest devices ($g_{m1} \gg g_{m3}..g_{m8}$) in order to disregard the noise contribution of the devices $M_3 - M_8$. Therefore a small (W/L) or a large $(V_{GS} - V_{th})$ in the devices $M_3 - M8$ (relative to the input devices M_1, M_2) must be used. Operating in the strong inversion region the transconductance of these devices shows a square root dependence on the drain current ($g_m \propto \sqrt{I_D}$) and hence it is smaller compared to the input devices which operate in the deep weak inversion reaching a high transconductance value. Care must be taken while sizing the transistors $M_3 - M5$. The second pole generated by the gate capacitances of these devices (C_{n1} and C_{n2} in Fig. 4.4) is sensitive for the stability of the OTA. Therefore the capacitance C_{n1} and C_{n2} must not grow large. A large load capacitance, C_L, at the output of the OTA stems the problem of the second pole. The used technology provide MIM capacitances on higher metal layer. A big C_L was placed on top of the active devices for best exploitation of the silicon area.

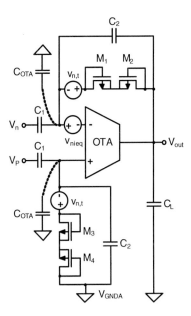

Figure 4.7: Preamplifier for invasive EMG-recording.

Fig. 4.7 shows a block diagram for the noise sources of the ac-coupled preamplifier. The input-referred noise of the preamplifier is calculated as a function of the dominated input-referred OTA noise (equation 4.28).

$$V_{ni,preamp}^2 = \left(\frac{C_1 + C_2 + C_{OTA}}{C_1} \right)^2 \cdot V_{nieq}^2 \qquad (4.31)$$

where C_{OTA} is the input capacitance of the OTA (gate capacitance of the input devices M_1 and M_2 of the OTA, Fig. 4.4). Larger input devices would increase the noise of the preamplifier. However the input devices must be sized large to reduce flicker noise, which is the main noise source in low frequency operation. A large C_1 would reduce the effect of C_{OTA} at the cost of silicon area. This might thus affect the input impedance of the amplifier as discussed later.

The power consumption and silicon area are highly dependent on the noise specification of the system. Therefore, an optimum is calculated using a noise specification value for the OTA based on [78]. The noise in electronic circuits is mainly affected by two values:

1. **Power consumption:** The relation between power consumption and noise can be illustrated on the example of the input-referred thermal noise of the OTA depicted in equation 4.29 as a simplification of the overall noise after reducing the flicker noise by increasing the area of the MOSFETs. Assuming $g_{m1} \gg g_{m3}, g_{m7}$ equation 4.29 is simplified to:

$$V^2_{\text{nieq,thermal}} = \frac{8kT\gamma}{g_{m1}}\Delta f \tag{4.32}$$

As the input devices work in weak inversion, the transconductance of M_1, M_2 is given by equation 4.5. 4.5 in 4.32 results in:

$$V^2_{\text{nieq,thermal}} = \frac{8kT\gamma}{\frac{I_D}{nU_T}}\Delta f = \frac{nU_T \cdot 8kT\gamma}{I_D}\Delta f \tag{4.33}$$

where $U_T = kT/q$ is the thermal voltage. The noise voltage is calculated by integrating equation 4.33 over the signal bandwidth Δf.

$$V_{\text{nieq,thermal}} = \sqrt{\frac{nU_T \cdot 8kT\gamma\Delta f}{I_D}} \tag{4.34}$$

The power consumption of the OTA is directly dependent on V_{dd} and the total current dissipated by the four branches of the OTA I_D. Using equation 4.34 it can be calculated as follows:

$$P_{\text{OTA}} = I_{\text{total}} \cdot V_{dd} = 4I_D \cdot V_{dd} = \frac{1}{V^2_{\text{nieq,thermal}}}4nU_T \cdot 8kT\gamma\Delta f V_{dd} \tag{4.35}$$

Equation 4.35 shows the relation between power consumption and the input-referred noise for MOSFETs operating in the weak inversion ($P \propto \frac{1}{V^2_{nieq,thermal}}$). In order to halve $V_{nieq,thermal}$ the power consumption must be multiplied by 4. Transistors operating in strong inversion show a square root dependence between the transconductance and the drain current, $g_m \propto \sqrt{I_D}$. Consequently, the power consumption scales as $1/V^4_{nieq,thermal}$ for MOSFETs operating in strong inversion compared to a $1/V^2_{nieq,thermal}$ for MOSFETs in the weak inversion.

2. **Area:** The area of the preamplifier is mainly set by the used capacitances which are placed in a higher metal layer above the active region (MOSFET region). The used technology offers a unit capacitance per area $C_{uc} \approx 1fF/\mu m^2$. The total area of the preamplifier is then calculated as:

$$\text{Area}_{\text{preamp}} = \frac{2C_1 + 2C_2 + 2C_L}{C_{uc}} = \frac{2C_2(A_{\text{preamp}} + 1) + 2C_L}{C_{uc}} \tag{4.36}$$

Using equation 4.57 for the higher cutoff frequency of the preamplifier and hence for the bandwidth of the preamplifier ($\Delta f \approx f_{LP} = \frac{G_{mOTA}}{2\pi C_L A_{preamp}}$) in equation 4.32 and integrating over Δf delivers:

$$V_{nieq,thermal} = \sqrt{\frac{8kT\gamma \cdot \Delta f}{g_{m1}}} = \sqrt{\frac{4kT\gamma}{\pi C_L A_{preamp}}} \qquad (4.37)$$

Substituting equation 4.37 in equation 4.36 results in:

$$Area_{preamp} = \frac{2C_2(A_{preamp}+1)}{C_{uc}} + \frac{8kT\gamma}{V^2_{nieq,thermal} \cdot \pi A_{preamp} \cdot C_{uc}} \qquad (4.38)$$

The area is inversely proportional to the square root of the thermal noise similar to the relation between power consumption and noise. An optimum gain can be calculated by choosing a minimal capacitance C_2 and the specified noise value. The minimum area for a noise value of 2 μV_{rms} was determined at $A_{preamp} \approx 50$ as depicted in Fig. 4.8.

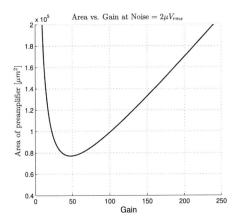

Figure 4.8: Area as a function of the gain of the preamplifier for a noise value of 2 μV_{rms}.

Feedback and Stability

Feedback is a way of combining the output of a system with its input. For example, in heating systems temperature of a room is the output of the system as well as an input parameter of the same system. The output temperature is subtracted

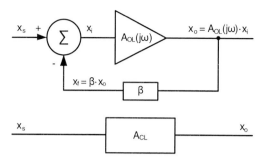

Figure 4.9: Block diagram of system feedback.

from the recorded temperature in order to regulate the room temperature to a desired value. Therefore, the feedback parameter is used to stabilize the system. In amplifiers a negative feedback is used to stabilize the amplifier circuit against changes in the device parameters such as process variations, V_{dd} variations, temperature, controlling input and output impedance levels, device aging. Two disadvantages arise when using negative feedback: reduced gain, and oscillation tendency [79]. To solve these problems further amplifier stages are used and extra attention for stability is required.

Fig. 4.9 illustrates the feedback principle in an amplifier system. A_{OL} represents the open-loop gain of the amplifier circuit and β is the feedback factor. The amplifier's input, x_i, is the difference between the system input, x_s, and the feedback signal, x_f:

$$x_i = x_s - x_f \qquad (4.39)$$

The output signal is represented by:

$$x_o = A_{OL} \cdot x_i = A_{OL} \cdot (x_s - x_f) \qquad (4.40)$$

Using ideal system components, the calculation of the closed-loop gain, A_{CL}, is given by:

$$A_{OL} = \frac{x_o}{x_i} \qquad (4.41)$$

$$\beta = \frac{x_f}{x_o} \qquad (4.42)$$

$$A_{CL} = \frac{x_o}{x_s} \qquad (4.43)$$

Applying $x_f = \beta \cdot x_o$ into equation 4.40 results in:

$$A_{CL} = \frac{x_o}{x_s} = \frac{A_{OL}}{1 + A_{OL} \cdot \beta} \tag{4.44}$$

Equation 4.44 illustrates the relationship between closed-loop and open-loop gain. For a large value of the open-loop gain ($A_{OL} \rightarrow \infty$), the closed-loop gain reaches $\frac{1}{\beta}$. A large A_{OL} is needed for an accurate gain setting by the feedback circuit.

For a feedback-gain $1/\beta = 50$ and an open-loop gain of $A_{OL} = 1000$, the closed-loop gain is 4.7% deviated from the desired value.

$$A_{CL} = \frac{1000}{1 + 1000/50} = 47.6 \tag{4.45}$$

The negative feedback might cause a phase shift in the input signal leading to a positive feedback which adds to the input signal. This results in instability of the system. The loop gain of an amplifier is defined as $T = A_{OL}\beta$. The system is stable when the phase of $A_{OL}\beta$ reaches $180°$ whereas the magnitude of the amplifier is well below 0 dB. The phase at 0 dB subtracted from the $180°$ is called the **phase margin (PM)** of the system. It is used to determine the stability of an amplifier circuit. A phase margin $> 45°$ is required for a stable system with small peaking and less oscillations.

The first pole of the OTA, f_{3dB}, was calculated in equation 4.14. The phase of the OTA magnitude experiences a phase shift of $-90°$ at f_{3dB}. The second pole of the OTA, f_{nd}, was determined by equation 4.15. Another phase shift of $-90°$ takes place at f_{nd} which leads to an overall phase shift of $-180°$. The second non-dominant pole must be placed sufficiently beyond the GBW to guarantee closed-loop stability. From equation 4.15 an expression for the phase margin can be determined.

$$\Delta\phi = -\tan^{-1}\left(\frac{\omega}{\omega_{nd}}\right) = -\tan^{-1}\left(\frac{2\pi f \cdot C_{n1}}{g_{m3}}\right) \tag{4.46}$$

As M_3 operates in strong inversion, equation 4.46 is simplified to

$$\tan(-\Delta\phi) = \frac{2\pi f \cdot C_{n1}}{\sqrt{\frac{W_3}{L_3} KP_3 I_d}} \tag{4.47}$$

Solving for W_3 we get

$$W_3 = \frac{L_3}{KP_3 I_d} \cdot \left(\frac{2\pi f \cdot C_{n1}}{\tan(-\Delta\phi)}\right)^2 \tag{4.48}$$

Equation 4.48 delivers an easy way for a first estimation of the width of the load transistors. Obviously, W_3 cannot be determined at once as the capacitance C_{n1} is dependent on W_3. Therefore an iterative sizing process is used for this purpose.

As the dominant pole induces a 90° phase shift for the unity gain frequency, f_{un}, the phase margin can then be calculated as:

$$\phi_M = 90 - \tan^{-1}\left(\frac{\omega_T}{\omega_{nd}}\right) - \tan^{-1}\left(\frac{\omega_T}{\omega_{3d}}\right) + \tan^{-1}\left(\frac{\omega_T}{\omega_{z1}}\right) \quad (4.49)$$

where $\omega_T = 2\pi f_T$ is the transition (unity-gain) frequency[4], ω_{nd} is the second pole ($\omega_{nd} = g_{m3}/C_{n1}$), ω_{3d} is the pole at node 4 (NMOS current mirror at Fig. 4.4) $\omega_{3d} = g_{m8}/C_{n4}$, and ω_{z1} is a zero at the same node 4 with $\omega_{z1} = 2\omega_{3d}$.

Design Approach

The design procedure of the OTA concentrated on noise reduction within low power conditions taking area constraints into consideration (mini ASIC: $1525x1525\ \mu m^2$). The transconductance of the input devices is determined using the cutoff frequency specification of the first order low-pass filter (section 4.2.2.2) for a given load capacitance C_L which was first determined through a maximum given area constraints for the preamplifier. An iterative procedure for optimizing the values of g_m and C_L was used to reduce the overall noise within the specified value (Table 4.1). The transconductance then determines the DC open-loop gain A_v together with the output resistance of the OTA. The input devices were only enlarged to reduce the flicker noise contribution. This did not greatly affect the transconductance of those devices (g_{m1}) as they operate in weak inversion and highly dependent on the drain current ($g_{m,weakinv} \propto I_d$). The transistors of the current mirrors have a large length to reduce their g_m and hence reduce their noise contribution as previously declared in the noise analysis section (see equation 4.28). Their area was enlarged by doubling width and length of the transistor (so it doesn't affect the transconductance g_m) to reduce flicker noise. The current mirrors are designed to place the parasitic poles and zeros (ω_{nd}, ω_{3d}, ω_{z1}) sufficiently beyond the transition frequency to guarantee closed-loop stability.

Fig. 4.11 shows the schematic of the preamplifier. It is widely used in electronic systems for medical applications [49], [80]. The preamplifier consists of an opera-

---

[4]The transition (unity-gain) frequency f_T is equal to GBW in a first-order system without parasitic poles or zeros.

tional transconductance amplifier (G_m OTA, see Fig. 4.4), capacitive feedback, and a MOS-bipolar transistors for the high pass corner frequency. The input capacitance C_1 is used for blocking DC-voltages (such as half-cell potential and body DC voltages) from saturating the preamplifier. High input impedance of the preamplifier compared to the electrode impedance is essential. The positive input of the preamplifier at V_p has the following impedance:

$$Z_{\text{in,p}} = Z_{C_1} + Z_{C_2||R_{\text{MOS}}}||Z_{\text{OTA,p}}, \ Z_{\text{OTA,p}} \approx C_{\text{OTA}} \approx C_{\text{gg,M1}}$$

$$= \frac{1}{j\omega C_1} + Z_{R_{\text{MOS}}||C_2||C_{\text{gg,M1}}}, \ C_{\text{gg2}} = C_2 + C_{\text{gg,M1}} \quad (4.50)$$

$$= \frac{1}{j\omega C_1} + \frac{1}{\frac{1}{R_{\text{MOS}}} + j\omega C_{\text{gg2}}}$$

where R_{MOS} is the resistance of the MOS bipolar element (several $G\Omega$), C_2 is the feedback capacitance ($C_2 = C_1/A_{preamp} = C_1/50$), and $Z_{OTA,p}$ is the input impedance of the OTA (high input impedance due to the input capacitance of the input devices $C_{gg,M1}$). The signal energy of the EMG lies generally in the 200 - 400 Hz region. At 200 Hz an electrode impedance of several tenths of kΩ (≈ 10 kΩ) exists [81]. The impedance of the preamplifier with $C_1 = 20$ pF, $C_{OTA} = 960$ fF, $C_2 = 400$ fF, and $R_{MOS} \approx 23$ GΩ reaches a value > 600 MΩ at 200 Hz (see Fig. 4.10). The differential input impedance ($\frac{1}{j\omega C_1}$, $C_1 = 20$ pF) is approximately ≈ 40 MΩ at 200 Hz which lies 4000 times bigger than the electrode impedance, ≈ 10 kΩ at 200 Hz.

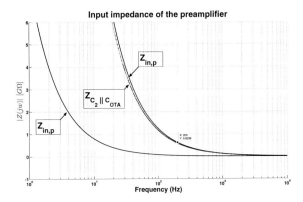

Figure 4.10: Input impedance of the preamplifier at the positive input V_p.

The diode-connected MOSFETs (PMOS devices) M_1, M_2, M_3, M_4, build a huge resistance parallel to the capacitance C_2. The function of these transistors is firstly to bias the input node of the OTA to virtual ground (V_{GNDA}). Secondly, it constructs the low cutoff frequency f_{HP} together with C_2. The resistance of the diode-connected transistors is dependent on the voltage across the transistor ($V_{GS} = V_{DS}$). V_{DS} changes its sign due to signal amplitude. Therefore, two symmetrical PMOS transistors are connected in series to reduce distortion caused by positive and negative signal amplitudes.

The input capacitance of the OTA is almost bigger than the feedback capacitance ($C_{OTA} > C_2$). Therefore, it must be considered in the calculation of the transfer function of the preamplifier. The transfer function of the amplifier is derived as follows:

1. Input V_n:

$$
\begin{aligned}
(V_n - V_-)sC_1 &= V_- sC_{OTA} + \frac{V_- - V_{out}}{R_{mos}||Z_{C_2}} \\
(V_n - V_-)sC_1 &= V_- sC_{OTA} + (V_- - V_{out})\frac{1 + sR_{mos}C_2}{R_{mos}} \\
V_n sC_1 + V_{out}\frac{1 + sR_{mos}C_2}{R_{mos}} &= V_-\left[sC_1 + sC_{OTA} + \frac{1 + sR_{mos}C_2}{R_{mos}}\right] \\
V_- &= \frac{V_n sR_{mos}C_1 + V_{out}(1 + sR_{mos}C_2)}{1 + sR_{mos}C_1 + sR_{mos}C_2 + sR_{mos}C_{OTA}}
\end{aligned}
\tag{4.51}
$$

2. Input V_p:

$$
\begin{aligned}
(V_p - V_+)sC_1 &= V_+ \frac{1 + sR_{mos}(C_2 + C_{OTA})}{R} \\
V_+ &= V_p \frac{sR_{mos}C_1}{1 + sR_{mos}C_1 + sR_{mos}C_2 + sR_{mos}C_{OTA}}
\end{aligned}
\tag{4.52}
$$

3. The output current i_{out} of the OTA flows mostly in C_L as $C_L \gg C_2$. This results in:

$$
V_{out} = (V_+ - V_-)\frac{g_m}{sC_L}
\tag{4.53}
$$

Using the three steps above, the transfer function is given as:

$$
H(s) = \frac{V_{out}}{V_p - V_n} = \frac{C_1}{C_2} \cdot \frac{1}{\frac{C_L}{g_m C_2}(sC_1 + sC_2 + sC_{OTA} + \frac{1}{R}) - \frac{1}{sR_{mos}C_2} - 1}
\tag{4.54}
$$

The simplified gain of the preamplifier is given by the capacitance ratio:

$$
A_{preamp} = \frac{C_1}{C_2}
\tag{4.55}
$$

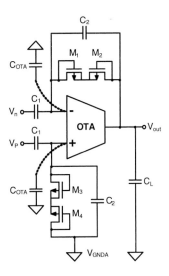

Figure 4.11: Schematic of the AC-coupled EMG preamplifier with capacitive feedback.

The bandwidth of the preamplifier ($BW = f_{HP} \cdots f_{LP}$) is determined using the diode-connected MOS-bipolar element R_{MOS} (M_1 and M_2 resistance in Fig. 4.11, see appendix B.3), the transconductance of the OTA (G_{mOTA}), the preamplifier gain (A_{preamp}), and the capacitances C_1, C_2, C_L.

$$f_{\mathrm{HP}} = \frac{1}{2\pi R_{\mathrm{MOS}} C_2} \tag{4.56}$$

$$f_{\mathrm{LP}} = \frac{G_{\mathrm{mOTA}}}{2\pi C_{\mathrm{L}} A_{\mathrm{preamp}}} \tag{4.57}$$

The low-pass filter of the preamplifier represents the first order of the high order analogue filter in the MyoC1-chip. Next section describes the analogue filter designed for biomedical application.

Layout

The preamplifier uses a symmetrical OTA for signal amplification. Because of its symmetry, the OTA suppose to deliver good CMMR results by rejecting common mode signals. Unfortunately, because of process and device mismatches the size of the transistors and hence their functionality is affected. Layout techniques are used to work against mismatches between active devices. A common centroid layout has been used in this design for the input devices as explained in Fig. 4.12.a. The large capacitances of the external circuitry of the amplifier have been layouted with the same technique (Fig. 4.12.b).

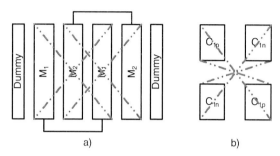

Figure 4.12: Common centroid layout structures.

4.2.2.2   Anti-Aliasing Filter

The low-pass anti-aliasing filter suppresses unwanted high frequency components of the input signal and prevents aliasing in the digital converted signal by attenuating high frequency components in the same signal (see chapter 2.2.2.2). The additional number of bits required by an ADC to accommodate out-of-band noise signals is [82]:

$$N_{add} = \frac{Int(\frac{I}{S} + 3)}{6 \text{ dB}} \tag{4.58}$$

where Int is an integer number, I is the power of out of band signals, S is the power of the signal of interest. The anti-aliasing filter improves the implementation and reduces the data rate of an ADC by attenuating the out-of-band signals (minimize I) and hence reducing N_{add}. This is especially important for an implantable application as wireless data transmission is the most expensive component in the system with respect to power and volume consumption.

In order to obtain an approximately brick-shaped transfer function and to avoid large overshoot of the step response, a Bessel low-pass filter was implemented [50], using the circuit shown in Fig. 4.13 [83]. A 5th order filter has been selected, which was realized by the contribution of the preamplifier (see above) and the series connection of two stages of the second order filter in Fig. 4.13. This filter order

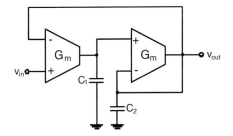

Figure 4.13: Structure of the second order $G_m C$-filter.

represents a compromise between sampling rate and analogue circuit resources, since a higher filter order allows a lower sampling rate and vice versa.

Several aspects have been considered prior to the implementation of the active filter. The design work was performed as follows:

1. Decide on the filter structure taking the following aspects into consideration: power consumption, tuning, frequency range, distortion, SNR, THD.

2. Pick a filter characteristically based on the needs of the high level application.

3. Calculate the transfer function of the filter.

4. Calculate the parameter of the transfer function using a mathematical tool (such as MATLAB).

5. Implement the operational amplifier for the filter using the specification parameter in the first step.

6. Simulate your filter to verify results.

The requirements for the anti-aliasing filter can be summarized as follows:

- Low power.

- Configurable cutoff frequency ($f_{3dB} = 800$ Hz $\|$ 1.5 kHz).

- Gain 0 dB.

- Input range < 600 mV.

- Total harmonic distortion < 0.1%.

- Noise: The input-referred noise content from the filter is divided by the gain of the preamplifier ($V_{in,filter} = V_{in,filter}/50$). This relieves the noise requirement for the analogue filter. However, noise power is an additive component and there are several components in the system (electrodes, preamplifier, filter, driver, post-amplifier, ADC) which deliver noise content to the overall system noise. Therefore noise reduction must take place in every design stage for a better overall noise performance of the system.

Filter Structure

There are several methods to implement an analogue filter:

- Active RC filter (Fig. 4.14a): The RC filter utilizes an operational amplifier[5] with resistors and capacitors in the feedback loop. Although high SNR values (100 dB) and low THD can be reached, it is hard to tune RC filters as the processing errors of capacitances and resistors can reach up to 20% of the desired value.

- MOSFET-C filter (Fig. 4.14b): The MOSFET device operating in linear region is used for the tuning of the time constant. An operational amplifier is therefore needed. Its external circuit is built using MOSFET devices instead of resistances. MOSFET-C filters are not suitable for large input signals because of linearity problems and large second order distortion [84].

- G_mC filter (Fig. 4.14c): The filter is built with an operational transconductance amplifier (OTA) acting as a resistance (G_m) followed by a capacitance. The transconductance of the OTA (G_m) can easily be tuned by changing the bias current of the OTA (G_m is a function of the bias current I_{bias}) to set up the time constant of the G_mC filter. A huge advantage of the G_mC filter beside tuning with I_{bias} is the low current consumption of the filter. Low G_m values or high capacitances are needed for low frequency operation. As the transconductance is proportional to the square root of the bias current in strong inversion or to the bias current in weak inversion, small G_m results in small bias current and hence in low power consumption. Distortion can be a problem for this filter which must be reduced by certain design strategies.

---

[5]The operational amplifier used for analogue filter might need to drive resistive loads which would increase the power consumption of the filter.

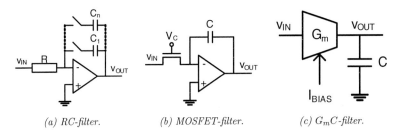

(a) RC-filter.   *(b) MOSFET-filter.*   *(c) G_mC-filter.*

Figure 4.14: Structures of active filters.

Table 4.3 shows a comparison between the three filter types mentioned above. G_mC filter fits best our application based on its low area, low power consumption, and its good tuning possibility. The disadvantages in distortion and dynamic range doesn't reach crucial values for the desired application.

Table 4.3: Comparison of filter structures.

Parameter	RC filter	MOSFET-C filter	G_mC filter
Power Consumption	high/medium	high/medium	Low
Tuning	hard	possible	good
Frequency range	low	low	high
Distortion	low	medium	medium
Dynamic range	high	medium	medium
Area	high	medium	low

Filter Characteristic

There are three most commonly used low-pass filter types: Butterworth, Tschebyscheff, and the Bessel filter. They mainly differ in their pass band, stop band, cutoff, and group delay[6] characteristics. The characteristics of these low-pass filter types are briefly described as follows:

---

[6] The group delay is the first derivative of the phase with respect to the angular frequency ($T_{gd} = -d\beta/d\omega$). It delivers information about the distortion in the signal. If T_{gd} is not constant over the signal bandwidth of a modulated signal, waveform distortion will take place [85]. It is necessary for good pulse transmission to have a flat group delay.

- **Butterworth low-pass filter:** The Butterworth low-pass filter characteristic was first described by the British engineer **Stephen Butterworth** in [86]. It is very flat in the pass band region. There is no gain ripple neither in the pass band nor in the stop band. Thus the cutoff behavior is relatively slow compared to the Tschebyscheff low-pass filter.

- **Tschebyscheff low-pass filter:** Tschebyscheff filter is named after the Russian mathematician Pafnuti Lwowitsch Tschebyschow. The Tschebyscheff low-pass filter has a steeper roll-off at the cutoff frequency than the Butterworth or the Bessel filter. However, it has ripples in the pass band region.

- **Bessel low-pass filter:** The Bessel filter is named after the German mathematician Friedrich Bessel. Its main characteristic is a maximal flat group delay and hence a linear phase response. Analogue Bessel filters maintain the wave shape of filtered signals in the pass band. They show no ripples in the frequency response. However, the roll-off steepness in the cutoff region has not been improved.

Table 4.4: Filter Characteristics.

Characteristic	Butterworth	Tschebyscheff	Bessel
Flatness of pass band	good	bad	moderate
Stop band attenuation	moderate	high	slow
Group delay flatness	ripple	poor	very flat
Ripple in pass band	no	yes	no

The characteristics of the three examined filters is illustrated in Fig. 4.15. A Bessel low-pass filter was chosen to best maintain the waveform of the filtered signals preventing step response overshoot (important for the acquisition of action potentials, see note 6) or distortion in the output signal.

Transfer Function and Filter Parameter

The transfer function of a low-pass filter is given by:

$$A(s) = \frac{A_0}{\prod_i^n (1 + a_i s + b_i s^2)} \tag{4.59}$$

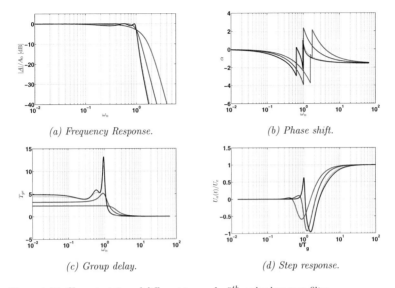

(a) Frequency Response.

(b) Phase shift.

(c) Group delay.

(d) Step response.

Figure 4.15: Characteristics of different types of a 5th order low-pass filter.
⎯⎯ Bessel, ⎯⎯ Butterworth, ⎯⎯ Tschebyscheff 0.5 dB, ⎯⎯ Tschebyscheff 1 dB.

where A_0 is the pass band gain, s is the complex frequency variable for the normalized form of the transfer function ($s = j\omega/\omega_C$), a_i and b_i are the filter coefficients, and n is the attenuation ratio above the corner frequency f_C with $-n \cdot 20$ $dB/decade$ (n is also called the filter order). For a first-order filter, b_1 is equal to zero resulting in:

$$A_1(s) = \frac{A_0}{1 + a_1 s} \tag{4.60}$$

A second order low-pass filter stage has the transfer function:

$$A_i(s) = \frac{A_0}{(1 + a_i s + b_i s^2)} \tag{4.61}$$

where i is the stage number of a combined high order low-pass filter. The coefficients for a 5th order low-pass filter are given in Table 4.5. The corner frequency f_c was specified to be $f_c = 800$ Hz or $f_c = 1500$ Hz. The overall transfer function of the anti-aliasing low-pass filter for a filter gain of $A_0 = 1$ is then calculated as:

$$A_5(s) = A_1 \cdot A_2 \cdot A_3 = \frac{1}{1 + a_1 s} \cdot \frac{1}{(1 + a_2 s + b_2 s^2)} \cdot \frac{1}{(1 + a_3 s + b_3 s^2)} \tag{4.62}$$

Table 4.5: 5th order Bessel filter coefficients [50].

i	a_i	b_i	f_{ci}/f_c	Q_i
1	0.6656	0.0000	1.502	-
2	1.1402	0.4128	1.184	0.56
3	0.6216	0.3245	2.138	0.92

Two low-pass filters have been implemented in the MyoC1-chip: 5th and 7th order low-pass filters (Fig. 4.16a and 4.16b). Signal quality and data rate reduction were the reason for testing a higher order filter at the expense of silicon area. Therefore, different types of capacitances have been used to implement high order low-pass filter within the given silicon area of the ASIC (1525 x 1525 μm²). The filter parameters (OTAs and capacitances) were designed for $f_c = 1500$ Hz. Switching to the lower corner frequency $f_c = 800$ Hz was accomplished with the bias current of the operational transconductance amplifiers as it directly regulates the transconductance G_m of the OTA. Only the capacitance of the first order low-pass filter was twice available for the lower corner frequency range $s_c = 800$ Hz. The first order of the low-pass filter was implemented at preamplifier stage. The load capacitance of the preamplifier C_L has been determined according to the given area for the preamplifier. The transconductance of the preamplifier is then calculated from the cutoff frequency f_{LP} (equation 4.57) of the first order low-pass filter given in Table 4.5.

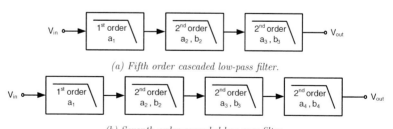

(a) Fifth order cascaded low-pass filter.

(b) Seventh order cascaded low-pass filter.

Figure 4.16: A 5th order and 7th order cascaded low-pass filter.

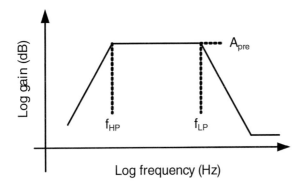

Figure 4.17: Plot of the gain versus frequency of the preamplifier.

$$f_{c1} = 1.502 \cdot f_c$$

$$f_{c1} = f_{LP} = \frac{G_{mOTA}}{2\pi C_L A_{preamp}} \qquad (4.63)$$

$$G_{mOTA} = 1.502 \cdot f_c \cdot 2\pi C_L A_{preamp}$$

where f_{c1} is the cutoff frequency of the first order filter (it is given in Table 4.5), C_L is the load capacitance determined by area constraints in the floor planning phase of the chip ($C_L = 40$ pF), A_{preamp} is the closed loop gain of the preamplifier ($A_{preamp} = 50$).

The second order stages of the anti-aliasing filter utilizes two OTAs and two capacitances as illustrated in Fig. 4.18a. The canonical topology of the second order low-pass filter was proposed in [87].

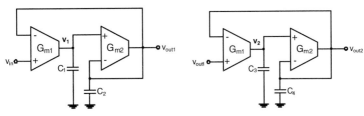

(a) *Stage one of a second order anti-aliasing G_mC filter.*

(b) *Stage two of a second order anti-aliasing G_mC filter.*

Figure 4.18: A canonical topology of two stages of the 5th order cascaded low-pass filter.

The calculation of the transconductance and the capacitances of the second order low-pass filter in Fig. 4.18a is given by

$$v_1 = \frac{G_{m1}}{j\omega C_1} \cdot \left(v_{in} - v_{out}\right) \tag{4.64}$$

$$v_{out} = \frac{G_{m2}}{j\omega C_2} \cdot \left(v_1 - v_{out}\right)$$

$$v_{out} = \frac{G_{m2}}{j\omega C_2} \cdot \left(\frac{G_{m1}}{j\omega C_1} \cdot \left(v_{in} - v_{out}\right) - v_{out}\right) \tag{4.65}$$

$$A(s) = \frac{v_{out}}{v_{in}} = \frac{1}{1 + \frac{j\omega C_1}{G_{m1}} + \frac{j\omega C_1 \cdot j\omega C_2}{G_{m1} \cdot G_{m2}}}, \ s_n = j\omega$$

$$A(s) = \frac{v_{out}}{v_{in}} = \frac{1}{1 + s_n \frac{C_1}{G_{m1}} + s_n^2 \frac{C_1 \cdot C_2}{G_{m1} \cdot G_{m2}}} \tag{4.66}$$

The transconductances, G_{m1} and G_{m2}, of the consecutive operational transconductance amplifiers are equally chosen to reduce design complexity of the OTAs ($G_{m1} = G_{m2}$). Using equation 4.61 and $A_0 = 1$ we get:

$$A(s) = \frac{1}{1 + s_n \frac{C_1}{G_m} + s_n^2 \frac{C_1 \cdot C_2}{G_m \cdot G_m}} = \frac{1}{(1 + a_1 s + b_2 s^2)} \tag{4.67}$$

As a result the parameters C_1, C_2, and G_m can be calculated using $s_n = w/w_c$ as follows:

$$\frac{a_1}{\omega_c} = \frac{C_1}{G_m} \tag{4.68}$$

$$C_1 = \frac{a_2 G_m}{\omega_c} \tag{4.69}$$

$$\frac{b_1}{\omega_c^2} = \frac{C_1 C_2}{G_m^2} \tag{4.70}$$

$$C_2 = \frac{b_1 G_m^2}{C_1 \omega_c^2} \tag{4.71}$$

where $s = j\omega/\omega_c$, and $\omega_c = 2\pi f_c$ is the cutoff angular frequency.

The calculation of the parameters for the second stage of second order low-pass filter in Fig. 4.18b is done identically to the first stage. The result is:

$$A(s) = \frac{1}{1 + s_n \frac{C_3}{G_m} + s_n^2 \frac{C_3 \cdot C_4}{G_m \cdot G_m}} = \frac{1}{(1 + a_3 s + b_3 s^2)} \tag{4.72}$$

As a result the parameters C_3, C_4, and G_m can be calculated using $s_n = w/w_c$ as follows:

$$\frac{a_3}{\omega_c} = \frac{C_3}{G_m} \tag{4.73}$$

$$C_3 = \frac{a_3 G_m}{\omega_c} \tag{4.74}$$

$$\frac{b_3}{\omega_c^2} = \frac{C_3 C_4}{G_m^2} \tag{4.75}$$

$$C_4 = \frac{b_3 G_m^2}{C_3 \omega_c^2} \tag{4.76}$$

where $s = j\omega/\omega_c$, and $\omega_c = 2\pi f_c$ is the cutoff angular frequency.

A circuit diagram of the preamplifier with a fifth order anti-aliasing filter is illustrated in Fig. 4.19. The 7th order low-pass filter has been designed using the cascaded topology in Fig. 4.16b. The first order was implemented in the preamplifier. An extra second order stage has been included to the fifth order filter to construct the seventh order low-pass filter. The transfer function is calculated similarly to the fifth order filter in equation 4.62 as follows:

$$A_7(s) = A_1 \cdot A_2 \cdot A_3 \cdot A_4 \tag{4.77}$$
$$= \frac{1}{1 + a_1 s} \cdot \frac{1}{(1 + a_2 s + b_2 s^2)} \cdot \frac{1}{(1 + a_3 s + b_3 s^2)} \cdot \frac{1}{(1 + a_4 s + b_4 s^2)}$$

The parameter of the filter were calculated using the Bessel filter coefficients in Table 4.6. The filter parameter (G_m and capacitances) can be calculated using equations

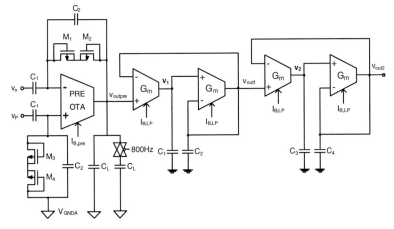

Figure 4.19: Diagram of the preamplifier with a 5th order anti-aliasing filter.

Table 4.6: 7th order Bessel filter coefficients [50].

i	a_i	b_i	f_{ci}/f_c	Q_i
1	0.5937	0.0000	1.684	-
2	1.0944	0.3395	1.207	0.53
3	0.8304	0.3011	1.695	0.66
4	0.4332	0.2381	2.731	1.13

$4.67 - 4.76$. The parameters for the extra stage of a second order low-pass filter are calculated as follows:

$$A(s) = \frac{1}{1 + s_n \frac{C_5}{G_m} + s_n^2 \frac{C_5 \cdot C_6}{G_m \cdot G_m}} = \frac{1}{(1 + a_4 s + b_4 s^2)} \tag{4.78}$$

As a result the parameters C_5, C_6, and G_m can be calculated using $s_n = w/w_c$ as follows:

$$\frac{a_4}{\omega_c} = \frac{C_5}{G_m} \tag{4.79}$$

$$C_5 = \frac{a_4 G_m}{\omega_c} \tag{4.80}$$

$$\frac{b_4}{\omega_c^2} = \frac{C_5 C_6}{G_m^2} \tag{4.81}$$

$$C_6 = \frac{b_4 G_m^2}{C_5 \omega_c^2} \tag{4.82}$$

where $s = j\omega/\omega_c$, and $\omega_c = 2\pi f_c$ is the cutoff angular frequency.

Design Strategy

The design of the filter was done using a trade-off between area and noise under the consideration of low power operation. The matlab software was used for the calculation of the filter parameters. A simple noise formula for the OTA noise was used to verify the selection of a transconductance of the OTA. The parameter selection for the anti-aliasing filter was done as follows:

1. Start with a low transconductance value ($G_m = 10$ nS).

2. Calculate the overall area of the capacitances[7] C_1, C_2, C_3, C_4.

---

[7]The area of a capacitance in the used technology was calculated as the value of the capacitance divided by the unit capacitance per square micrometer ($A_{capacitance} = C/A_{1\mu m}$).

3. If the calculated area of the filter capacitances doesn't fit into the available filter area on ASIC, increase G_m and step back.

4. Compare the noise value for the chosen G_m and the calculated capacitances with the specified filter noise value (the filter is allowed to deliver more noise than the preamplifier because of the gain factor of the preamplifier which reduces the input-referred noise of further stages in the analogue channel. A reasonable threshold for the filter noise is $V_{ni,filter} < A_{preamp} \cdot V_{ni,preamp}/2$). If the filter noise exceeds the threshold, increase G_m to reduce the noise and go back two steps.

5. Terminate.

OTA Implementation

The filter OTA has the same topology as the preamplifier (Fig. 4.4). The differential pair of the symmetrical OTA operates in weak inversion. The transconductance G_m of the amplifier correlates with the bias current (see equation 4.5). For low G_m values extremely low currents in the nano- or picoampere region can be used. Furthermore, the current mirror factor $B = 0.5$ has been chosen to reduce the copied current and hence reduce the G_m of the OTA at the expense of noise (which is less critical in this stage). The cutoff frequency of the OTA $f_c = \frac{G_m(I_b)}{2\pi C}$ is controlled by the bias current. Low current value is needed in order to save extra capacitance area for the low frequency application. For the higher cutoff frequency ($f_c = 1.5$ kHz) the bias current was chosen to be $I_{Bias} = 20$ nA. To switch to the lower cutoff frequency ($f_c = 800$ Hz) the bias current was set to half of its initial value ($I_{Bias} = 10$ nA). Such a low bias current causes the transistors to operate in weak inversion. A long length has been chosen for all the transistors in order to keep them in saturation. Unusual long length devices made the layout of the filter OTA a challenge. It was not possible to use common centroid method for the layout. Therefore, special care was given to symmetry.

The noise modeling of the filter OTA is similar to the modeling of the noise of the preamplifier's OTA. Based on the derivation of equation 4.28 for the OTA of the preamplifier and considering the current mirror factor (B = 0.5), we can calculate the overall simplified noise power of the filter OTA as follows:

$$I_{n,out}^2 = B^2(2I_1^2 + 2I_3^2) + 2I_3^2 + 2I_7^2 = \frac{I_1^2}{2} + \frac{5I_3^2}{2} + 2I_7^2 \quad (4.83)$$

$$V_{nieq,LPOTA}^2 = \frac{I_{n,out}^2}{g_{m1}^2} = V_{n1}^2 \left[\frac{1}{2} + \frac{5}{2} \cdot \left(\frac{V_{n3}^2}{V_{n1}^2} \right) \left(\frac{g_{m3}}{g_{m1}} \right)^2 + 2 \left(\frac{V_{n7}^2}{V_{n1}^2} \right) \left(\frac{g_{m7}}{g_{m1}} \right)^2 \right] \quad (4.84)$$

Flicker noise of the filter OTA is the biggest component in the input-referred overall filter noise. Increasing the gate area of the devices has been used to reduce flicker noise to an acceptable level.

4.2.2.3   Driver

The OTA of the anti-aliasing filter uses 20 nA bias current for the higher bandwidth 1500 Hz and 10 nA for the lower bandwidth of 800 Hz. With such a low current the OTA possesses a low driver capability and hence it is not able to drive the input impedance of the post-amplifier. Therefore a driver stage is needed in order to deliver more current at the output stage to quickly drive the gate of the post-amplifier in a multiplexed application. An operational transconductance amplifier has been chosen for this stage because it got current at its output and it is able to drive capacitive load (the input of the post-amplifier is the gate capacitance of its input devices). The driver amplifier has the specification parameter in Table 4.7. Eight analogue components follow the preamplifier in the analogue channel (preamplifier, six OTAs for the low-pass filter 7th order, and the post-amplifier). All of them ought to deliver the following amount of noise $V_{noutx} = 50 \cdot V_{ni,preamp}/8$ as their input-referred noise components would be divided by the preamplifiers gain ($A_{preamp} = 50$). As a result, the overall input-referred noise of the analogue channel is equal to: $2 \cdot V_{ni,preamp}$. The design procedure for the driver OTA follows the same steps as that of the OTA of the preamplifier.

4.2.2.4   Multiplexer

The MyoC1 chip has five analogue channels connected to one post-amplifier through a multiplexer (Fig. 4.1). A block diagram of the multiplexer is depicted in Fig. 4.20. The multiplexer is controlled from outside of the chip using three digital input signals (s_1, s_2, s_3) for the five analogue channels ($\log_2(5) = 2.3Bit \rightarrow 3Bit$). The external control signals are demultiplexed internally in the digital demultiplexer and one channel is then activated according to the digital control value (see the timing

Table 4.7: Specification parameter of the driver OTA.

Parameter	Value
V_{dd}	1.2 V
I_{Bias}	$< 2 \mu$ A
Temperature	$37°C$
Closed loop gain	0 dB
Closed loop bandwidth	> 1500 Hz
Max input/output range	600 mV
Input-referred noise	$< 12 \mu V_{rms}$
Phase margin	$> 45°$

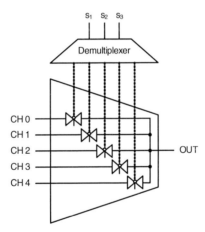

Figure 4.20: Block diagram of the multiplexer on the MyoC1-chip.

diagram in Fig. 4.21).

Switching the analogue channels is accomplished through a transmission gate controlled by the demultiplexer[8]. The schematic of the transmission gate is depicted in Fig. 4.22. The width of the MOSFETs used in the transmission gate affects the channel resistance and hence the propagation delay through the transmission gate. However, increasing the width of the MOSFETs enlarges the gate area (C_{oxn}, C_{oxp})

---

[8]A transmission gate is used because NMOS passes well a zero and a PMOS passes well the one. As a result, a rail-to-rail output swing is possible [36].

and thus increases the load for the control signals. The transmission gate is designed
to work at the highest possible switching frequency for the design ($f_{clk} = 1.4$ MHz
$\rightarrow 100$ KS/s $= 1$ Mbps at the output of the ADC). A dummy switch circuit is used
to reduce the effect of charge injection and clock feedthrough [36]. The drain and
source of the dummy device are shorted. Its control signal is complement of the signal controlling the devices beside. Its control signal is slightly delayed using double
the width of the neighbored devices (higher gate capacitance leads to higher delay).

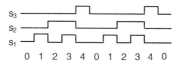

Figure 4.21: Timing diagram of the multiplexer's control signals.

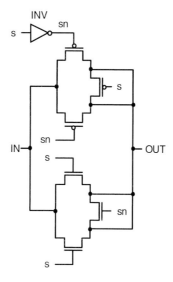

Figure 4.22: Schematic of the transmission gate for the multiplexer.

4.2.2.5   Post-amplifier

A rail-to-rail programmable post-amplifier of Tomasik et al. [88] has been used
to further amplify the recorded signals and to drive the input capacitance of the
ADC within the time limits of the ADC clock. The noise performance of the post-
amplifier is less critical at this stage based on the gain of the preamplifier (Fig. 4.2).
The post-amplifier is built in a non-inverted topology as depicted in Fig. 4.23. The
amplification factors are configured through the transmission gates $TG_0 - TG_3$. The
following resistances have been used:

$$R_{1a} = 140 \text{ k}\Omega, \text{ } R_{1b} = 80 \text{ k}\Omega, \text{ } R_{1c} = R_2 = 10 \text{ k}\Omega \tag{4.85}$$

The following gain factors have been accomplished: $A_{postamp} = 1$, 2.4, 12, 24. The
gain of 1 is configured in reset state as the control signal S_0 is complemented per
default. Table 4.8 lists the control signals and their associated gain factors.

Table 4.8: Amplification factors of the post-amplifier.

S_3	S_2	S_1	S_0	Gain
0	0	0	0	1
0	0	1	1	2.4
0	1	0	1	12
1	0	0	1	24

The noise of the post-amplifier is controlled through two bias currents which
depict a trade-off between power consumption and noise. Furthermore the input-
referred noise for all bias current configurations[9] was $< 2.2 \text{ } \mu V_{rms}$.

The post-amplifier operates in a multiplexed application [48] [89]. Several channels are switched on/off at its input. In order to load the input capacitance of the
ADC (≈ 70 pF) within the time limits of the ADC[10], the post-amplifier must work
at least in the medium range ($I_{ab} = 15 \text{ } \mu A$, $I_{diff} = 93 \text{ } \mu A$). The equivalent circuit in
Fig. 4.24 is used to determine the needed power consumption of the post-amplifier
to load the ADC capacitance with an error less than 1/2 LSB. At the beginning

---

[9]There are two bias currents available for the post-amplifier. They range from high pow-
er/low noise to low power/high noise operation. $I_{ab} = 30 \text{ } \mu A \cdots 15 \text{ } \mu A \cdots 1 \text{ } \mu A$ and $I_{diff} = 143 \text{ } \mu A \cdots 93 \text{ } \mu A \cdots 13 \text{ } \mu A$.

[10]The switch at the input of the ADC is closed for $2.5 \cdot f_{clk}$ of the ADC clock signal.

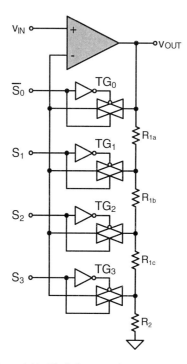

Figure 4.23: Block diagram of the post-amplifier.

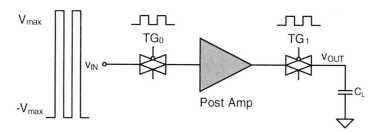

Figure 4.24: Equivalent circuit for simulating the needed power consumption of the post-amplifier.

of the simulation the input signal v_{IN} is applied then the transmission gate TG_0 was closed (TG_1 is open). The amplified input signal $A_{post} \cdot v_{IN}$ then emerges at the output of the post-amplifier. Three microseconds after closing TG_0 the second transmission gate TG_1 is then closed. The charging process of the load capacitance C_L starts immediately after closing TG_1. The time at which the load capacitance C_L is charged within an error of 1/2 LSB is then determined and listed in Table 4.9.

Table 4.9: *Timing values to load the ADC input capacitance within an error of $< 1/2LSB$.*

Gain		1		2.4		12		24	
Edge[11]@ low operation	r	f	r	f	r	f	r	f
Settling time[12][μs]	15	10.7	> 25	7.73	> 25	> 25	> 25	> 25
Edge @ medium operation	r	f	r	f	r	f	r	f
Settling time [μs]	0.76	0.46	1.81	2.36	1.73	1.74	2.2	1.31
Edge @ high operation	r	f	r	f	r	f	r	f
Settling time [μs]	0.51	0.23	1.2	2.25	1.26	1.2	0.83	0.18

4.2.3 Analogue to Digital Converter (ADC)

EMG signals in digital form are necessary for wireless signal transmission and further signal processing. Digital signal processing is cheaper than analogue processing in terms of power consumption and technological possibilities. Thus, it is important to convert the analogue recorded muscle signals into digital form for processing and transmission purposes. An analogue-to-digital converter (ADC) is needed for digitizing analogue signals. ADCs differ in their characteristics. For the underlying implantable EMG recording system there are several important requirements on the ADC. In addition to resolution and sampling rate the ability to deal with multiplexed inputs is necessary for the ADC. Different types of ADC architectures exist. Section 2.2.2.3 includes an aggregation of the most famous ones.

A brief comparison of the introduced architectures in section 2.2.2.3 favors the SAR type of ADC. A sigma-delta converter requires indeed no anti-aliasing filter as its sampling rate is much higher than the effective rate and because of the digital filtering at the output. However, the sigma-delta modulator is slow concerning effective data rate and less useful in a multiplexed application. Pipeline converters

---

[11]r: rising edge, f: falling edge
[12]Time needed for an error $< 1/2$LSB

have high speed and moderate bit resolution upto 16 bit. But they show some latency to get the sample at the output. Furthermore, a pipeline ADC may require more power for less latency and it consumes more silicon area than the SAR converter. The Flash ADC is the fastest architecture available. Though it consumes more power than a SAR converter and has small bit resolution as the number of comparators increases by a factor of two for every extra bit and must be more accurate. However, SAR converter would need more accurate components for higher bit resolution.

A successive approximation register (SAR) ADC is used in MyoC1 chip. SAR suits the multiplexed application best and it fits well into the intended application in terms of sampling rate and resolution (Fig. 2.19). Furthermore, SAR architecture fits into a small silicon area and requires low power [90]. The user application can reduce the resolution by picking the needed number of bits from the parallel output of the SAR. This is used later on in order to adjust the data rate of the wireless interface according to the available link quality as implantable devices have dynamical surroundings, that continuously change their electrical properties based on the ionic movement in body fluids.

4.2.3.1   Bit Resolution

The resolution of the ADC is a key factor prior to design and during specification phase. The area and power consumption are directly affected by the bit resolution. Therefore an exact specification of the bit resolution is required to our implantable application in order to save area and reduce the power consumption which are critical parameters for implantable applications [91] [92].

Bit Resolution vs. Area

As seen in Fig. 4.28 the capacitance value increases with the exponent of 2 for every additional bit ($2^N \cdot C$) in the SAR architecture. Therefore, the area of a SAR-ADC correlates exponentially with the bit resolution. The unit capacitance in the MyoC1 chip is: $C_u = 63.3$ fF. The area of two unit capacitances[13] including wiring is approximately $A_{2Cu} = 11.45~\mu m \cdot 22.4~\mu m = 256.48~\mu m^2$. For a 10 bit SAR-ADC the required unit capacitances equal $2^N = 2^{10} = 1024$. As a result, the overall area of the ADC is approximately $A_{ADC} = A_{2Cu}/2 \cdot 2^N = 131317~\mu m^2$. The area

---

[13]The capacitances of the ADC were designed as MIM-type capacitances (metal-insulator-metal).

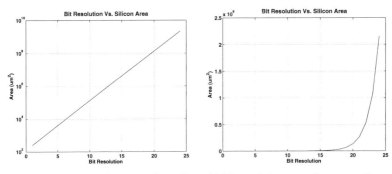

(a) Bit resolution vs. silicon area (logarithmic view).

(b) Bit resolution vs. silicon area (linear view).

Figure 4.25: Bit resolution vs. silicon area.

dependency upon the bit resolution is depicted in Fig. 4.25 for a unit capacitance of $C_u = 63.3$ fF.

Bit Resolution vs. Power

The power consumption of the SAR architecture correlates exponentially as well with the bit resolution. As the energy on a capacitor is $E_C = \frac{1}{2}CV^2$, a maximum power consumption is obtained for $V = V_{dd}$. The overall capacitance of the ADC multiplies duplicates for every additional bit $C_{ADC} = 2^N \cdot C_u$ (C_u is the unit capacitance in Fig. 4.28). A 10 bit SAR converter consumes an energy of approximately:

$$E = \frac{1}{2} 2^N C_u V_{dd}^2$$
$$E_{10} = \frac{1}{2} 2^{10} \cdot 63.3\text{fF} \cdot 1.2^2 = 46.4\text{pJ} \tag{4.86}$$

where $C_u = 63.3$ fF is the unit capacitance, $V_{dd} = 1.2$ V is the supply voltage (maximum voltage at the input is less or equal V_{dd}). The SAR converter of the MyoC1 chip needs 14 cycles to convert one sample. The maximum sampling frequency of the MyoC1-chip is calculated as follows:

$$F_{ADC} = F_{nyq} \cdot R_o \cdot K \cdot 14 = 3.36 \text{ MHz} \tag{4.87}$$

where $F_{nyq} = 2 \cdot BW = 2 \cdot 1500$ Hz = 3 kHz is the Nyquist frequency, $R_o = 16$ is an oversampling factor for anti-aliasing, and K = 5 is the maximum number of active channels.

The power consumption is defined as the energy per time. As a result the power is calculated as:

$$P = \frac{E}{T} \tag{4.88}$$

where E is the energy, and T is the duration. From $T = 1/F_{ADC}$ we obtain for the power of a 10 bit ADC $P_{10} = E_{10} \cdot F_{ADC} = 46.4 \; pJ \cdot 3.36 \; MHz = 156.6 \; \mu W$. The power consumption upon the bit resolution is depicted in Fig. 4.26 for a unit capacitance of $C_u = 63.3$ fF and a supply voltage of $V_{dd} = 1.2$ V.

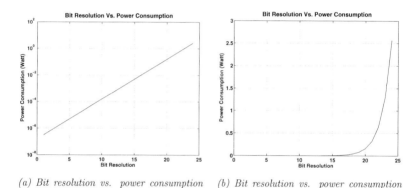

(a) Bit resolution vs. power consumption (logarithmic view).

(b) Bit resolution vs. power consumption (linear view).

Figure 4.26: Bit resolution vs. power consumption.

Bit Resolution vs. SNR

An in-vitro measurement of internal EMG showed that bio-signals in the range of $2 \; mV_{pp}$ can be recorded with the implanted electrodes. No experimental electrode-noise measurement has been issued for the epimysial electrodes described in section 3.1. A root-mean-square noise voltages from $1 \; \mu V_{rms}$ to $15 \; \mu V_{rms}$ were identified at the body surface using Ag-AgCl electrodes [73]. Assuming an input noise (electrode noise) of 10 μV_{rms} the SNR of the input signal is calculated as:

$$V_{n,\text{electrode}} = \sqrt{2} \cdot V_{n,\text{electrode,rms}} = 14.14 \ \mu V \tag{4.89}$$

$$\text{SNR}_{\text{input}} = 20 \log \frac{A_{\text{signal}}}{A_{\text{noise}}} = 20 \log \frac{1\text{mV}}{14.14 \ \mu V} \approx 37 \ \text{dB} \tag{4.90}$$

$$\text{SNR}_{\text{ideal}} = 6.02N + 1.76 \Rightarrow N = 5.85 \Rightarrow N = 6 \ \text{Bit} \tag{4.91}$$

where N is the bit resolution.

A higher resolution than the one calculated in equation 4.91 would cost more resources (power and area) at the ADC. However, the implantable electrodes are in a continuous development process concerning impedance reduction and stability which gives hope for better noise performance. The additive electronic noise from the analogue front end can be reduced at the cost of power and area. The resolution was specified to be 10 bit after considering the former aspects. The SAR architecture provides additionally the opportunity for the external application to reduce the bit resolution if needed. This can be done by cutting the LSBs of the SAR output parallel register which are located below noise level.

Data Rate

The ADC is used in a multiplexed application. Therefore, its sampling rate is calculated as the sampling rate of a single channel multiplied by the number of analogue channels. The sampling rate of analogue channels has been chosen with consideration to aliasing, area, and power consumption. The Nyquist condition for sampling analogue signals generally requires $f_{sample} \geq 2 \cdot f_{signal}$. The maximum power of unwanted fractions in the recorded signal has been specified so that the error is smaller than 1/2 LSB. Consequently, we obtain:

$$\frac{V_{\text{Err}}}{V_{\text{LSB}}} = \frac{1}{2} \frac{1}{2^N} = \frac{1}{2} \frac{1}{2^{10}} = \frac{1}{2^{11}} \tag{4.92}$$

$$20 \log_{10} \frac{V_{\text{Err}}}{V_{\text{LSB}}} = -66 \ \text{dB} \tag{4.93}$$

where N = 10 is the bit resolution of the ADC. The minimum sampling frequency of the analogue channel (Fig. 4.27) according to the Nyquist criterion is then calculated as:

$$f_{\text{ch,sample}} = 2 \cdot f_{-66\text{dB}} \approx 48 \ \text{kHz} \tag{4.94}$$

The minimum required data rate for the ADC is then given as:

$$f_{\text{ADC,sample}} = N \cdot f_{\text{ch,sample}} N_{\text{ch}} = 10 \ \text{Bit} \cdot 48 \ \text{kHz} \cdot 5 \ \text{Channels} = 2.4 \ \text{Mbps} \tag{4.95}$$

where N is the bit resolution of the ADC, $f_{ch,sample}$ is the sample rate per channel, and N_{ch} is the number of analogue channels at the input of the ADC.

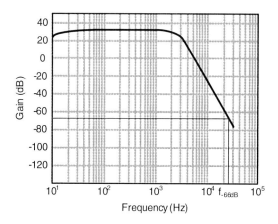

Figure 4.27: Diagram of the frequency response of the analogue channel with 5th order anti-aliasing filter.

4.2.3.2   Architecture Considerations

The architecture of the SAR converter for the MyoC1 chip is illustrated in Fig. 4.28. It is based on the work of W. Galjan in [48] [89]. The SAR-ADC consists of a capacitor array (C, $\cdots$, 512C), a comparator, and a switching circuit including a shift register for the 10 Bit digital output. Two preamplifiers are used in front of the comparator (SW_AMP) in order to increase accuracy by amplifying the input amplitude. A dynamic latch follows the preamplifier (Comparator) which works in low power and has low offset voltage. The SAR switching circuit implements the successive approximation algorithm by successively switching the bottom plates of the capacitor array starting from MSB till the LSB. The digital output of the SAR is saved in a 10 Bit shift register. An end of conversion (nEOC) signal is generated while saving the digitized value in the 10 Bit register. The ADC can be powered down with the Start signal for saving power in case it is not used. The SAR-ADC requires 14 clock cycles per sample (10 bit output data). Fig. 4.29 depicts the timing of the ADC for one sampling period.

4.2.4 Digital Control Register

The configuration register is 40 bits long and stores the parameters for the control of the chip. It is realized as a shift register (Fig. 4.30) that can be read from or written to by using a serial interface. The configuration register controls the following parameters:

- Signal bandwidth: The bandwidth of the analogue channels can be selected

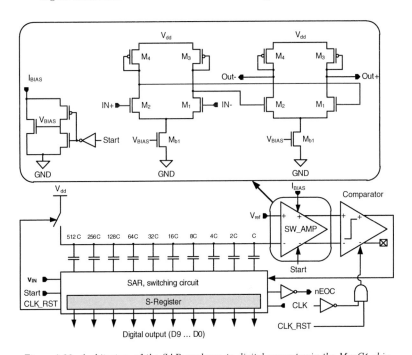

Figure 4.28: Architecture of the SAR analogue-to-digital converter in the MyoC1 chip.

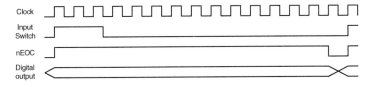

Figure 4.29: Timing diagram of the SAR-ADC for sampling 10 bit.

between (100-800 Hz) and (100-1500 Hz). It is defined by a capacitor array in
parallel with the load capacitance of the preamplifier and a proper selection
of the bias currents of the filter OTAs.

- Channel 4: This channel has a multiplexer at its input for higher reliability
  in case one electrode pair at the input is defect. An electrode pair with the
  lower impedance value will be selected (external control).

- Amplification factor of the post-amplifier: The gain of the post-amplifier is
  configured using four bits of the configuration register for controlling the transmission gates of the post-amplifier according to Table 4.8.

- Power-Down: If necessary each channel can be powered down individually.

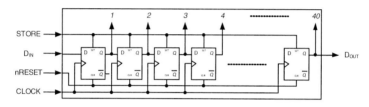

Figure 4.30: 40 bit configuration register.

4.2.5   Chip Architecture

Fig. 4.31 shows the architecture of the MyoC1 chip for invasive EMG recording.
The chip has five analogue channels each of which consists of a preamplifier, a high
order anti-aliasing filter, and a driver OTA. The analogue channels are connected
through an analogue multiplexer to one post-amplifier followed by a 10 bit SAR-
ADC. A low-noise, low-power preamplifier is used to amplify the signal difference of
extracellular potentials (20 μV_{pp} to 2 mV_{pp}) between two closely located epimysial
electrodes in a bipolar configuration (see chapter 2.2.1). The preamplifier consists
of an AC-coupled operational transconductance amplifier (OTA) with a capacitive
feedback. The AC-coupling is used to protect the preamplifier from being saturated
by the body DC voltage at the recording sites and the electrode half-cell potential.
The feedback network (parallel RC) is composed of a capacitance and a MOS bipolar
element to accomplish a high resistance for a high time constant.

Current prosthetic applications utilize surface EMG between 100 - 500 Hz. The high-pass cutoff frequency of the preamplifier is designed to be $f_{HP} = \frac{1}{2\pi R_{MOS}C_2} = 7$ Hz and the low-pass cutoff frequency is selectable (800 Hz, 1500 Hz) in order to examine low frequency and high frequency components in the invasive recorded EMG respectively. A high order anti-aliasing filter is implemented to reduce data rate and reject unwanted high frequency components in the recorded signal. A G_mC type of filter is used to easily switch between two cutoff frequencies and for its low power consumption. The anti-aliasing filter possesses a Bessel characteristic with flat group delay to improve quality and reduce distortion in the output signal. A driver OTA follows the filter to drive the multiplexer and hence the input impedance of the post-amplifier. The analogue channels are then multiplexed by the analogue

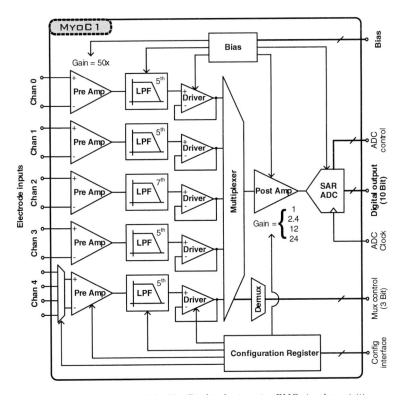

Figure 4.31: Architecture of the MyoC1 chip for invasive EMG signal acquisition.

multiplexer placed at the output of the driver.

The area of the analogue components of the analogue channel is determined by the area of the used capacitances. Different types of capacitances were implemented to reduce the silicon area and hence to examine the usability of these capacitances. The implemented capacitances are as follows:

- Metal-insulator-metal (MIM): The MIM capacitance is the most stable one concerning voltage dependency and process deviation.

- Metal-oxide-metal (MOM): MOM capacitors are implemented on metal layers using the lateral capacitive coupling between the plates formed by metal lines [93] and using vias to different metal layers (Fig. 4.32a). The MOM capacitor achieves higher-unit capacitance-value than the MIM capacitance because more metal layers can be used.

- MOS capacitance: The MOS capacitance is located between the gate and the shorted regions of drain, source, and substrate (Fig. 4.32b). The MOSFET capacitance C_{ox} is the highest among the used capacitances. However, it suffers from voltage dependency which makes its usability critical for high AC voltages.

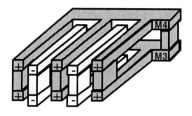
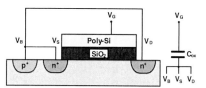

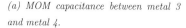

(a) MOM capacitance between metal 3 and metal 4.

(b) MOSFET capacitance between the gate and the shorted net drain-source-substrate.

Figure 4.32: Capacitance types.

The capacitances of the analogue components of the chip are summarized in Table 4.10.

---

[14]First order of the anti-aliasing filter is designed using an OTA. The preamplifier is built using an amplifier the same as the post-amplifier.

Table 4.10: Capacitances used in the analogue section of the MyoC1 chip.

Channel	Preamplifier	Preamplifier Load	Filter	Filter order
0	MIM	MOS, MOS	MIM	5
1	MIM	MIM, MOM	MIM	5
2	MIM	MOS, MOS	MOS	7
3	MIM	MOS[14]	MIM	5
4	MIM	MIM, MOM	MIM	5

The multiplexer is followed by a programmable class AB post-amplifier for further amplification (R2R output) of the recorded EMG signals. The post-amplifier is responsible for driving the input capacitance of the ADC within time conditions. The power consumption of the post-amplifier, 1.4 mW, is the largest among all components of the chip ($\approx$ 78% of overall power consumption). A successive-approximation-register (SAR) ADC is chosen to digitize the multiplexed five analogue channels because of its low power consumption [94], low silicon area and its suitability for multiplexed application. The bit resolution was designed to be 10 bit based on the SNR of the EMG signal at the input (see equation 4.91) and the fact that fewer bits can be used from the external application. Extra resolution would cost more area (the area of capacitances increases exponentially for every extra bit, $2^N C$) and hence more power (power consumption increases for the ADC and for the post-amplifier to load the enlarged input capacitance of the ADC). The sample rate of the ADC is $5 \cdot 48$ kHz $= 240$ kS/s. The sample rate is directly controlled by the clock frequency of the ADC. Reduced rates can be configured by setting lower ADC clock frequency[15].

A 40 bit configuration register is used to configure the bandwidth of the analogue components, the input multiplexer of channel 4, the amplification factors for the post-amplifier, and the power-down functionality for the analogue components on the chip.

---

[15]The data rate of the chip must be configurable in order to assimilate the bandwidth of the wireless link.

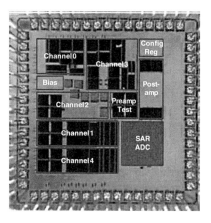

Figure 4.33: Fabricated MyoC1 chip.

4.2.6 Measurement Results (MyoC1)

The MyoC1-IC was fabricated in a 130nm 1P8M2T mixed-mode CMOS process. The size of the chip is $1525x1525 \ \mu m^2$ (Fig. 4.33). The measured power consumption of the chip is 1.8 mW (1.5 mA @ 1.2 V) when all channels are activated and the ADC clock is 1.4MHz (sample rate = 100 kS/s). The digital part (ADC digital control block and the configuration register) consumes a negligible $\approx 17 \ \mu W$ (14 μA @ 1.2 V).

The output of the post-amplifier and the input of the ADC were specially routed to external pads. They were shorted externally for measurement or recording purposes. Therefore, a separate measurement of the ADC was possible and is depicted in Table 4.11. For the measurement a sinus signal was used with the shape of:

$$s = 1.2V \cdot \sin(2\pi \cdot \frac{1}{150\text{Hz}}) \tag{4.96}$$

The effective number of bits (ENOB) is calculated by applying a ramp signal at

Table 4.11: SAR ADC measurement results.

SINAD [dB]	SNR [dB]	THD[16] [dB]	SFDR [dB]	ENOB (bits)
55.27	55.36	-70.62	71.95	8.88

---

[16]FThe THD is calculated by using the harmonics from the 2nd to the 5th one.

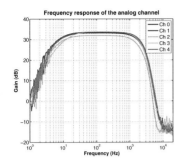 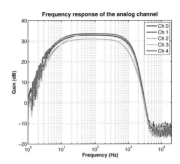

(a) Frequency response of the analogue channels for $\Delta f = 1500$ Hz.

(b) Frequency response of the analogue channels for $\Delta f = 800$ Hz.

Figure 4.34: Measurement results of the frequency response of the analogue channels.

the input of the ADC and then calculating the residuals of a linear ramp from the measurement curve out of the ADC. The maximum difference (per bits) is subtracted from the bit resolution and results in the ENOB of the ADC.

Fig. 4.34 shows the frequency response of the analogue channels for the two bandwidths (800 Hz and 1500 Hz).

Table 4.12 summarizes the measurement results for the MyoC1 chip. The MyoC1 chip fulfilled the specification parameters in Table 4.1. Furthermore, a reasonable value for CMRR has been accomplished due to the symmetrical structure of the used OTA. The power consumption per channel is given as 360 μW. Approximately 78% of the power is dissipated by the post-amplifier.

The needed current to load the input capacitance of the ADC is calculated by considering the output resistance of the post-amplifier and the input capacitance of the ADC. The maximum needed current is dissipated at t = 0 as follows:

$$V_0 = V_R + V_C = I \cdot R + \frac{Q}{C}$$
$$I_{max} = \frac{V_0}{R} \tag{4.97}$$

The voltage on the ADC capacitance V_C must be charged to within $\frac{1}{2}$LSB of the output voltage of the post-amplifier for an accurate 10-bit conversion. Following equation can be used to calculate the minimum sampling time for a 10-bit conversion:

$$t_s > R_{opost} \cdot C_{ADC} \cdot \ln(2^{11}) \tag{4.98}$$

where t_s is the sampling time, R_{opost} is the output resistance of the post-amplifier, and C_{ADC} is the input capacitance of the ADC.

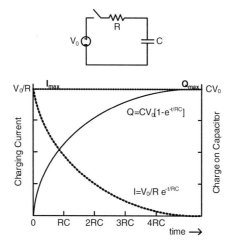

Figure 4.35: Capacitance current and charge.

The power consumption can be reduced by putting less effort on noise reduction in the design of the post-amplifier as explained in section 4.2.1 and reducing the input capacitance of the ADC[17].

---

[17]The input capacitance of the ADC can be reduced by internally connecting the post-amplifiers output to the input of the ADC and leaving aside the connection to an external pad which possesses a capacitance of couple of pico-Farads.

Table 4.12: MyoC1 measurement results.

Parameter	Value	Unit	Comment
Number of channels	5		6th multiplexed channel
Input range	±0.5 ±1 ±5 ±12	mV	Externally configurable
Channel gain	50 (preamp) 120 600 1200		Externally configurable
Bandwidth	$7-800$ $7-1500$	Hz	Externally configurable
Input-referred noise	2.17 2.4	μV_{rms}	$7-800$ Hz $7-1500$ Hz
CMRR	> 69	dB	@ 50 Hz
ADC resolution	10	Bit	
ADC THD	-70	dB	2nd - 5th harmonic
ADC ENOB	8.88	Bit	
Chip area	$1525x1525$	μm^2	Mini ASIC
Chip power consumption	1.8	mW	1.5 mA @ 1.2 V

Chapter 5

Wireless Telemetry

Modern implantable devices utilize wireless approach for power and data transmission. The usage of wires penetrating the skin for power or rather data transmission limits the movement of the subject and exposes infection risk. Therefore, a transcutaneous telemetry is used for power and data transmission. Two different frequencies are used for power and data transmission. Power is transferred using the 125 kHz frequency, and data is transmitted using the MICS band at 402 MHz. For high data rate a high carrier frequency is needed. Although the transmitted power is proportional to the frequency, the energy drops fast whenever the receiver moves out of tuning (caused by the dynamical surroundings with variable parasitics due to ionic fluid movement). Therefore, a low frequency is used for power transmission with wide range for matching imperfections. High frequency is used for data transmission which needs less energy for transmitting data and offers high data rate.

5.1 Wireless Power Transmission

A wireless powered system offers a huge advantage for an implantable biomedical system by saving battery usage and reducing surgical intervention at the cost of losses in energy transmission. The wireless power-system consists of an external class-E power amplifier (PA), an inductive pair of coils (external and internal), a resonant circuit, a rectifier, and voltage regulators on implant electronic system. The schematic of the power amplifier is depicted in Fig. 5.1. It utilizes an active device (NMOS) functioning as a switch, a load-network consisting of a high reactance V_{DD} shunt-feed choke (L_{choke}) and an output capacitance C_1 and a series-connected

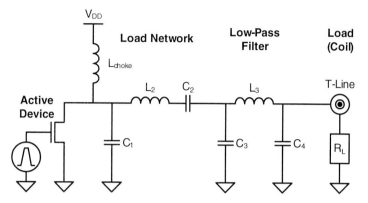

Figure 5.1: Class-E amplifier with an output low-pass filter.

LC element [95]. At the output there is a 50 Ω impedance of a coaxial cable.
Therefore, the output must match the 50 Ω impedance of the coaxial cable for
maximum power transmission. A low-pass filter is used to suppress harmonics of the
switching frequency. The parameter of the power amplifier is calculated according
to the formulas in [96]. The class-E amplifier might reach high efficiency values
by minimizing switching losses at the transistor though the efficiency of the power
transmission system is not the main goal at this stage. Large fluctuations in the
distance and the angle between the external and internal coil limit the efficiency of
the energy transmission system. Therefore, we designed the system for delivering the
needed amount of energy for a coil-to-coil distance of up to $8cm$, and an angle change
of up to 45° between the internal and external coil to guarantee an uninterrupted
recording. Due to the conductivity of biological tissue [97], a magnetic field is used
instead of an electric one for energy transmission to prevent power dissipation by
the tissue. The power-transmitting antenna is designed according to the Biot-Savart
law:

$$\mathrm{B} = \frac{\mu_0}{4\pi} \oint \frac{\mathrm{Id}\ell \times \hat{\mathrm{r}}}{|\mathrm{r}^2|} \tag{5.1}$$

where B is the magnetic field, μ_0 is the magnetic constant, r is the displacement
vector to the target point, $\hat{r}$ is the unit vector of r. The field along the z-direction

(centered to the loop) is calculated as:

$$dB_z = \frac{\mu_0 I_L d\ell}{4\pi} \frac{r}{(z^2 + r^2)^{3/2}}$$

$$B(z) = \frac{\mu_0}{4\pi} \frac{2\pi r^2 I_L}{(z^2 + r^2)^{3/2}} \quad (5.2)$$

For an N-turn wire loop (inductance L) the magnetic field is then calculated as:

$$B(z) = \frac{\mu_0}{4\pi} \frac{2\pi r^2 N \cdot I}{(z^2 + r^2)^{3/2}} \quad (5.3)$$

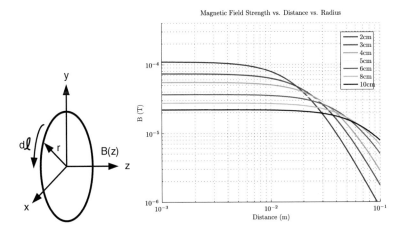

Figure 5.2: *Magnetic field strength from current in a wire loop.*

The equivalent circuit of the external coil is depicted in Fig. 5.3 as RLC circuit along with the equivalent circuit on the implant side. The impedance of the antenna is calculated as:

$$Z(j\omega) = R_{par} + j\omega L_1 + \frac{1}{j\omega C_1} \quad (5.4)$$

where R_{par} is the parasitic resistance of the loop antenna, L_1 is its inductance, and C_1 is the resonance capacitance. At resonance the angular frequency ω is equal to $\omega_0 = \frac{1}{\sqrt{LC}}$. The impedance at resonance is then calculated as:

$$Z(j\omega_0) = R_{par} + j\omega L (1 + \frac{1}{j^2\omega^2 LC}) = R_{par} \quad (5.5)$$

The current into the antenna at resonance is derived as [98]:

$$I(j\omega_0) = \sqrt{\frac{P_{in}}{N \cdot R_0}} \quad (5.6)$$

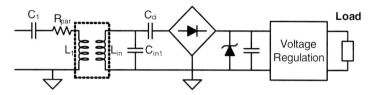

Figure 5.3: Schematic of the transmitting antenna and the receiver system.

where N is the number of turns, and R_0 is the parasitic resistance per loop. Substituting 5.6 into 5.3 yields:

$$B(z) = \frac{\mu_0}{4\pi} \frac{2\pi r^2}{(z^2 + r^2)^{3/2}} \sqrt{\frac{P_{in} N}{R_0}} \qquad (5.7)$$

As a result, the magnetic field generated by the external antenna is directly controlled by the number of turns and the primary input power (square root relation). Increasing the number of turns N for more magnetic field has a practical limit due to the increase of resistive parasitics and proximity effect [98]. An external wire loop with N = 14 turns and 8 cm radius oscillating at 125 kHz was able to supply the implant with 45 mW through the specified distance.

It is not easy to choose a transmission frequency for wireless power transmission. Many factors must be considered in advance. High frequency allows for small reactive components on implant module and hence saves area. Though, as the energy antenna is wrapped around the implant electronic board, high frequencies induce high eddy currents (power lost due to eddy current $P \propto f^2$) on the copper planes in the implant board and hence introduce a risk for heat dissipation and consequently affect surrounding tissue. Furthermore, parasitic components from the body tissue or fluids affect the matching of the antenna and hence may bring the antenna out of resonance easily and therefore reduce its quality. A lower frequency has a wide matching range and offers more flexibility concerning parasitic effects and introduces fewer eddy currents. Therefore, a 125 kHz transmission frequency was chosen for wireless power transmission.

Fig. 5.3 depicts a schematic of the power-transmission antennae and the internal receiver circuit. The internal coil on the implant module surrounds the implant printed circuit board (PCB) to fulfill the volume specification of the implant system as shown in Fig. 5.4 (coin 2 €). It encloses the implant and hence saves extra area and volume though the implant components lie in the magnetic field of the

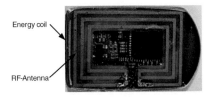

Figure 5.4: Implant electronic surrounded by the energy coil.

external coil. The main concern using this approach is the coupling effect from the energy interface into the analogue part of the system (into the analogue inputs of the ASIC "MyoC1"). This is considered as an extra noise source for the acquisition system. A high coupled voltage might saturate the preamplifiers of the MyoC1-chip. Therefore, a passive RC-filter was used at the analogue inputs of the MyoC1-chip to reduce the effect of the coupled voltage from the energy transmission system.

The received energy on the implant side is pretty much dependent on the volume of the secondary coil. The volume specifications are given through the implant size and the packaging thickness (1 cm) by the doctors. Therefore, there is no possibility for optimization by increasing the volume. The main factor to control the coupled energy at secondary side is by the magnetic field strength of the primary antenna.

5.2 Wireless Data Transmission

A medical implantable RF transceiver [70] was used for wireless data transfer. The transceiver fulfills the regulatory requirements for the MICS band. Furthermore, it consumes low power and has a small package which makes it suitable for implantable application. The maximum reached data rate was approximately 250 kbps. Only four channels were used during the experiment. As a result, we were able to record maximally $\approx$ 214 kbps using 10 bit resolution and a sampling rate of 5.3 kSps per channel. A printed board antenna was designed for the MICS band (402 - 405 MHz) [99]. The efficiency of the antenna in real application circumstances allowed to record a maximum data rate of $\approx$ 142.8 kbps.

Chapter 6

Implant System Design

The overall goal of the implant system (MyoImp) is to record internal EMG during targeted muscle contraction for investigating signal quality and usability in prosthetic application versus state of the art control mechanisms using surface recorded EMG. A short term implantation period was planned to investigate the challenges of such implantation and reveal the difficulties behind it. Therefore, a design concept is needed to make the system work reliably. A well-defined specification based on professional knowledge in several fields is inevitable.

Implant system design requires the collaboration (teamwork) between physicians and engineers to outline and discuss the condition boundaries of the system implantation. The physicians provide the dimension parameters for the implant module and the external boundary conditions for it using their biological knowledge and expertise. The engineers, on the other hand, propose a design methodology and concentrate on how to design the implant system in respect to the external boundary conditions. The following sections describe the design strategy of the implant system. It concludes information about the specification for the implant electronic system such as geometry (volume of the implant) and implantation strategy defined by the doctors. Furthermore, it describes the architecture of the implant electronic system through several development steps and the proposed implant prototypes.

Fig. 6.1 depicts the architecture and functional diagram of the implant system. It consists mainly of external and internal sub systems.

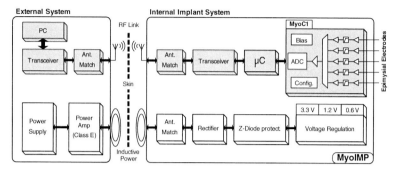

Figure 6.1: Architecture and functional block diagram of the implant electronic system.

6.1   Specification of the Implant System

Several aspects must be addressed prior to the design of a functional implant system and some of them emerge at first during live measurements. The geometry of the implant is specified based on the implant location inside the body. Therefore, physicians provide anatomical information for the designer to consider during conception. Furthermore, the distance between internal modules and external counterparts is partially specified by medical staff and animal trainer. The following itemization presents the most important specification parameters for the implant electronic system within animal experiments:

- **Geometry:** The area of the implant module was specified to be as big as a 2 Euro coin[1]. The thickness is not supposed to be more than 1 *cm* in order to avoid skin irritation and obviate the animal from scratching the skin at the implant location.

- **Packaging:** The implantation period usually defines the type of packaging of an implantable device. For long-term packaging a metal (titanium-alloy packaging used in pacemakers and cochlear implants), ceramic, or glass package [100] is used to hermetically seal the implant electronics. For short-term implantation though, silicone package is mostly enough to protect the electronics for several months. The package material affects the wireless interfaces of the

---

[1]After the first implantation, physicians stated that a bigger area is possible because of the high tolerance of the implant module by the rhesus macaque.

implant module extensively. Especially metal packages affect the energy and power transmission conception of the implant system. The coil and the antenna allocate extra volume outside the implant case.

A silicone package has been used for the animal experiments due to the short implantation period and the geometrical flexibility of the package.

- **Energy and data transmission:** The energy coil has been wrapped around the implant board in order to reduce the overall thickness of the implanted modules and for simplicity reasons. However, the implant electronics would be in the magnetic field of the external energy coil. Therefore, eddy currents are provoked on the board. The implant system design must take these effects into consideration as explained later in this chapter. The measurement setup directly affects the distance and angle between internal and external coil. The distance was specified to be ≈ 7 *cm* and the angle might reach $\approx 45°$.

  The data antenna (PCB antenna) was placed on top of the implant board. A tuning routine has been implemented on the implants microcontroller in order to tune the antenna which considers the parasitic effects from the surrounding tissue and from body fluids.

- **Number of EMG recording sites:** Four different muscles have been targeted for the rhesus macaque experiment: biceps, triceps, deltoid (front), deltoid (back). The muscles were chosen to reflect the targeted hand movement during task assignments in the measurement setup. Although multiple electrodes on one muscle would offer the opportunity for investigating the selectivity of invasive recording for prosthetic control, state of the art algorithms for prosthetic control were supposed to be tested using invasive recording. Therefore, different muscle groups were chosen for implantation.

6.2 Implant Electronic Module

High integration density of electronic components is quite important in implantable devices. In medical applications it is necessary to increase functionality on relatively small volume in order to reduce the dimension of the implanted device. The system measurement in the laboratory is not enough in most cases to get an implant system to function reliably in the measurement room after implantation. Several problems emerge at first during live measurement. Therefore, the implant electronic-system has undergone several development updates to reliably acquire invasive EMG data. This section presents the development process of several implant modules and the needed optimization steps (in the software and in the hardware) during the development.

6.2.1 MyoImp1 (Prototype)

The first implant electronic module (MyoImp1) was developed as proof-of-concept [4]. The main goals for this development step are listed as follows:

- **Area specification:** The initial area specification of a 2 euro coin was most important.

- **Number of channels:** Four channels in bipolar configuration were connected.

- **Packaging:** A metal case (Kovar) was intended to be used for implantation.

- **Energy coil:** The coil is wrapped around the metal package.

- **Data antenna:** A monopolar wire antenna $\left(\frac{\lambda}{4}\right)$ has been tested for data transmission at MICS band $\approx 402\ MHz$. A 50 Ω monopolar antenna has a length of:

$$\ell = \frac{\lambda}{4} \tag{6.1}$$

$$\lambda = \frac{c}{f_0} \tag{6.2}$$

$$v = \frac{c}{\sqrt{\mu_r \epsilon_r}} \tag{6.3}$$

where λ is the wavelength, $c = 3 \cdot 10^8\ m/s$ is the light velocity, μ_r is the relative permeability of the tissue (≈ 1 in the body), and ϵ_r is the relative permittivity of the tissue. Table 6.1 summarizes the dielectric properties (based on the work of IFAC [6]) of body tissues at $402\ MHz$ needed for the calculation of the monopolar wire length.

- **Digital processing unit:** An Atmel microcontroller (ATMEGA 88PA at 1 MHz) is responsible for collecting the data from the ASIC and for forwarding it to the wireless transceiver for wireless data transmission. No signal processing is implemented on the microcontroller.

Table 6.1: *Dielectric properties of body tissues at 402 MHz [6].*

Tissue	Conductivity [S/m]	Relative Permittivity	Loss Tangent	Wave length [m]	Penetration Depth [m]
Air	0	1	0	0.7458	N/A
Blood	1.3503	64.16	0.94104	0.08547	0.034304
Fat	0.041151	5.5789	0.32982	0.31163	0.30872
Muscle	0.79682	57.112	0.62386	0.09454	0.05255
SkinDry	0.68892	46.741	0.65907	0.10406	0.055225
SkinWet	0.66967	49.865	0.60051	0.10147	0.058262

Fig. 6.2 depicts the first prototype of the implant electronic board [4]. The ASIC was bonded on the board for best exploitation of the board area. Digital and analogue regions were spatially separated on board to avoid coupling between them as seen in the bottom view. The implant board was implemented in four metal

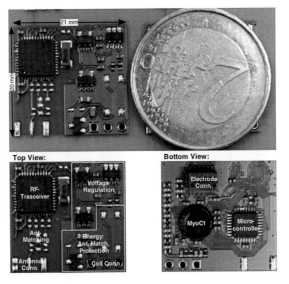

Figure 6.2: *Prototype I: Implant board for proof of concept [4].*

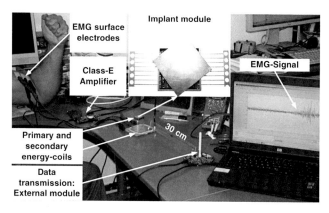

Figure 6.3: Surface EMG recording using prototype I.

layers (The top and bottom layers were used for signals. Whereas two internal layers were used for power and ground). Two internal copper planes (gnd and pwr) further separate and shield the analogue block from the high frequency RF-transceiver.

The MyoImp1 module was placed inside a Kovar package. Electrode inputs, coil, and antenna pins were connected to the glass-sealed leads of the Kovar package. The secondary coil was wrapped around the Kovar package and bipolar connected to the implant board through the package leads. A $\frac{\lambda}{4}$ wire monopolar antenna has been used to transmit data on the MICS frequency. A class-E amplifier is used for power supply to the primary energy coil which bridges a distance of 7 cm towards the implant coil. A data rate of 140 kbps has been accomplished in the air over a distance of 30 cm as depicted in Fig. 6.3.

The MyoImp1 module has been tested in continuous operation with a wireless data rate of $\approx$ 114 kbps[2] for several hours. The overall power consumption of the module is 10.2 mA at 3 V voltage supply. The current consumption of the individual system components is subdivided as follows:

- 5.7 mA in the RF-transceiver;

- 4.5 mA in the microcontroller, ASIC, and in the voltage regulators.

---

[2]The data rate of the MyoC1 ASIC is equal to the ADC clock frequency divided by 14 cycles for every sample. The RF-transceiver feeds the ASIC with the clock signal for the ADC. A clock frequency of 200 KHz leads to a sampling frequency of $f_s = 200 \ kHz/14 \approx 14.3 \ kHz$.

A flexible electrode structure using polyimide substrate and platinum contacts [5] was tested in an animal experiment (in a rhesus macaque). An evaluation study showed suitability for EMG recording [101]. The electrode structures were placed epimysially on the hand muscles of the Musculus deltoideus and routed to a connector attached to the head of the animal. A recording session using the MyoImp1 module and the implanted electrodes is depicted in Fig. 6.4. The measurement verified the usability of the developed implant electronic system for internal EMG acquisition.

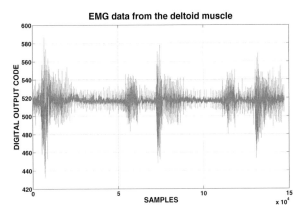

Figure 6.4: Subcutaneously recorded EMG from the deltoid muscle of a rhesus macaque using epimysial tape electrodes [5].

6.2.2 MyoIMP2: Implantation

The second prototype of the implant electronic board has been developed with regard to an implantation into a rhesus macaque. The implantation goal was to verify the overall functionality of the implant system including the epimysial electrodes and implant electronic module. The form and the dimension of the implant module has been specified with respect to the anatomy of the implant placement in the body. Physicians proposed to place the implant caudal between the scapulae of a primate monkey (rhesus macaque). Therefore, an elongated shape has been proposed (Fig. 6.5). The implant electronic board has a round shape at one side to avoid sharp angles and hence tissue damage due to continuous movement of the surrounding

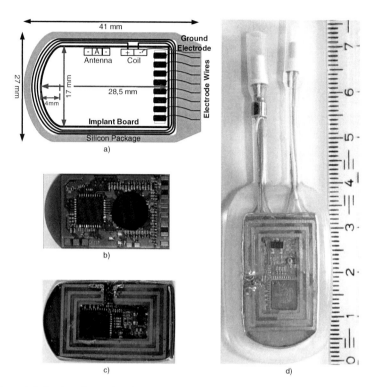

Figure 6.5: Implant design for the first animal experiment in a rhesus macaque: a) design block diagram, b) implant electronic board, c) energy coil and antenna assembly, and d) silicone packaged implant module.

muscles. Furthermore, the angles of the board have been arced to avoid any damage of the wound energy coil due to external pressure on the packaged implant. In order to reduce the overall thickness of the packaged implant module, a thin implant electronic board ($d = 0.5$ mm) has been used for the implantation. The silicone packaging enclosed the implant within $\approx 2 - 4$ mm of silicone. The electrode wires were routed from the implant electronic board through two strands to two 6 position male Nano-Miniature circular connector of Omnetics. A metal ring has been used to short the ground of the implant electronic board to the surrounding tissue in order to avoid ground loops (R_{gnd} in Fig. 6.6 is the resistance between two different grounds where current, I_{gnd}, might flow). This might introduce ripple voltage on the

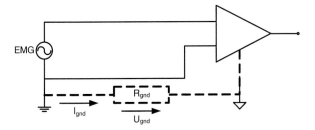

Figure 6.6: Ground loop in an amplifier system.

recorded EMG signal [102] and hence corrupt the signal acquisition or even saturate the input amplifiers.

The electronic components on the MyoImp2 board are the same as on the MyoImp1 except the microcontroller. A Texas Instruments low-power microprocessor from the MSP430 family is used instead of the ATMEGA 88PA for collecting the 10 bit digital EMG data from the MyoC1 chip and forwarding the data to the RF-transceiver for wireless transmission. Furthermore, it implements a protocol for the bidirectional communication with an external subsystem connected to a PC (base station). Thereby it is possible to externally configure the conditioning circuit for optimized operation and control the output data rate of the system to assimilate it to the available bandwidth of the RF-transceiver. The MSP430 microcontroller has more processing power than the Atmel counterpart for slightly more power consumption. Furthermore, a higher number of pins introduces more control possibilities. The MyoC1 chip was directly bonded on the implant electronic board which reduces the overall area of the board.

The MyoImp2 is capable of capturing the temperature locally at the board through the MSP430 microcontroller. The measured temperature was integrated into the communication protocol between the implant and the external base station. Hereby it is possible to monitor the increase in temperature due to eddy currents from the primary energy coil and hence to adjust the external energy supply to the needed minimum.

The energy coil has been wrapped around the implant board using a special epoxy isolated wire suitable for energy transmission using 125 kHz frequency in terms of skin effect. The distance between the external and internal energy coils has been specified to be 7 cm based on the measurement chair. A PCB data antenna for the

MICS band [99] was assembled on top of the implant board.

Test cases were planned to verify robustness of the implant against mechanical pressure and temperature effects. The packaged module was successfully tested in a salin solution within temperature range from $30°C - 60°C$. The data antenna showed a high sensitivity to the movement of the electrode wires in the solution and the orientation of the antenna is essential for uninterrupted loss-less wireless data transmission.

The MyoImp2 module was implanted in a rhesus macaque in the German Primate Center (DPZ) to verify functionality of the whole implant system including the epimysial electrodes. It was possible to successfully acquire internal EMG signals from the internal epimysial tape electrodes. Fig. 6.7 depicts an internal recorded EMG signal from the deltoid muscle and its power spectrum.

EMG signal transmission was no longer possible several weeks after implantation. During a live investigation of the reason behind the malfunction, wireless communication was established. The communication between the microcontroller and the ASIC seemed not to work properly (possible reason: loose contact between the on board bonded ASIC and the microcontroller). Explantation was therefore necessary in order to further investigate the implant electronic board. The inspection of the implanted system revealed several important issues:

- Ring (ground) electrode: The ground electrode is a metallic ring. A wire is welded to the metal ring and routed to the ground pin on the implant electronic board. As silicone does not adhere on metal, the interface between metal and silicone possesses a risk for body fluids to diffuse into the board. The pressure from body muscles on the implant branches containing the ring electrode would result in continuous movement of the branches and hence would enlarge a cleft at the interface between silicone and metal which in turn would accelerate the diffusion of body fluid into the implant board (Fig. 6.8).

- Deposition of haloid substance: A haloid substance has been sighted on several spots on the implant board (Fig. 6.9). It's origin is not clear yet.

- Flexible electrodes (tape or stripe electrodes): As mentioned in chapter 3.1, the bonding of the wire to the stripe electrode [5] contact was vulnerable to strong muscle movement. This led to a break in the bonding plate and thus to lose contact with the electrode sites.

Based on the inspection results, a redesign of the implant system was necessary. Stripe electrodes have been exchanged with silicone electrodes which possess more robustness and higher mechanical stability (chapter 3.1). The ring electrode was removed and one silicone electrode with shorted two sites has been used to connect the ground to the body tissue. Fig. 6.10 depicts the redesigned implant module MyoImp2-B.

MyoImp2-B was then implanted again in the rhesus macaque caudal between the scapulae. The silicone electrodes were placed epimysially on the following muscles: a) biceps brachii b) triceps surae c) deltoid clavicularis d) deltoid acromialis. A silicone electrode has been placed near the implant module to connect the ground of

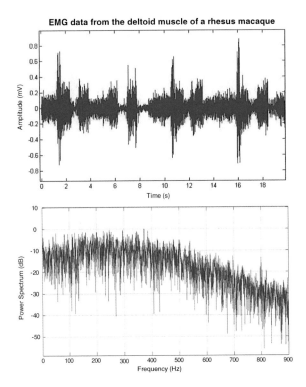

Figure 6.7: Internal EMG recording from the deltoid muscle using the MyoImp2 and its power spectrum.

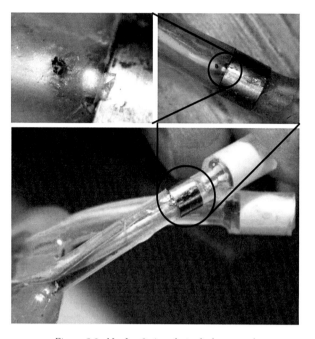

Figure 6.8: MyoImp2 ring electrode for ground.

the chip to the body tissue. The electrodes were inserted longitudinally under the epimysium of the target muscle and fixated using two sutures over the epimysium. Electrode wires were tunneled towards the back below the right scapula. The implant was inserted into the muscle and fixated under the muscle. The functionality of the implant module has been tested during the surgical operation by stimulating the muscle tissue near the implanted electrodes and detecting the stimulation artifacts using the MyoImp2-B module. The implantation went successfully. One week after the implantation it was possible to make the first recording. Reliable energy transmission depicts a real challenge because of the strong movement of the monkey. The angle and the distance between the energy coils was bigger than specified. It was possible to bridge the distance by using more energy on the primary coil which reduces the efficiency of the energy transmission. The efficiency was of less importance at this stage as the distance and angle specification for human implantation are totally different.

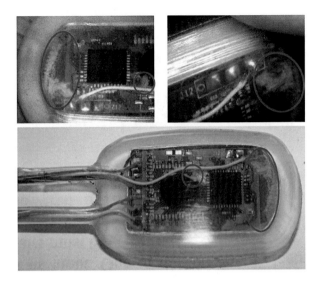

Figure 6.9: Haloid deposition on the MyoImp2 board.

The measurement setup for the EMG recording sessions uses a touch screen in front of the primate and the monkey must click on the lightened circle on the screen in order to be rewarded with juice or water. The trained primate repeated the assignments in several directions on the touch screen starting from the center as depicted in Fig. 6.11 [103]. Several synchronization signals were recorded simultaneously to the internal EMG acquisition (Fig. 6.11 top). A trigger signal is common between the measurement system and the implant system to align the synchronization signals and the EMG together on the same time axes. Hereafter it is possible to verify the EMG feature extraction algorithms. The recorded EMG showed high similarity of movements in the same direction whereas it showed clear differences between different movements as discussed in [81].

The functionality of the implant system has been successfully verified through long-term recording sessions (> one hour). Data has been gathered from four implanted electrodes over more than six months. However, artifacts such as those explained in section 6.3.2 were detected on the recorded signals. Fig. 6.12 depicts the recorded signals from the target muscles. Channel 4 shows the aforementioned artifacts.

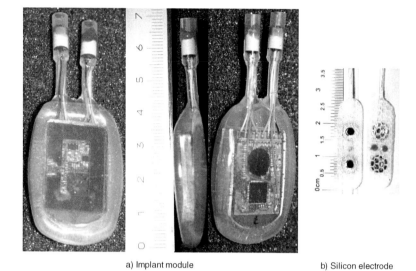

a) Implant module   b) Silicon electrode

Figure 6.10: MyoImp2-B: Redesign of the MyoImp2 implant module without ground ring and the implantable silicone electrode.

6.2.3   MyoIMP3

The functionality of the MyoImp2-B module was verified during the second implantation in a primate. However, several aspects were still challenging and require further development and optimization. The following issues were addressed during a third development series to improve functionality of the implant system:

- High RF noise signals contaminated the RF data transmission link and hence increased the link vulnerability. This is why the measurement room was shielded with aluminum foil during the measurement with the MyoIMP2-B. As a result the room temperature increased and affected the motivation of the primate to execute hand movement tasks. A technical solution on the implant board was needed, though.

- The thin implant board was vulnerable for mechanical stress from surrounding tissue. Assembling the small electronic elements and bonding the ASIC on board was technically possible but the mechanical stability of the board is not guaranteed. The silicone package doesn't protect the implant board from

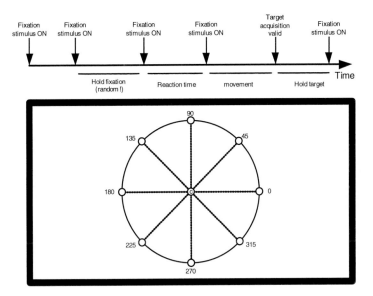

Figure 6.11: Targets on a touch screen for the primate to touch during measurement sessions.

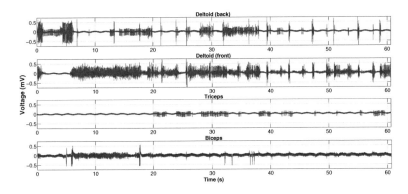

Figure 6.12: EMG recording using the MyoImp2-B module.

mechanical stress (bending of the packaged board due to muscle contraction). Therefore, during the redesign of the implant electronic system, a board thickness of 1 mm has been implemented to increase mechanical stability and hence offer better protection for rigid circuit components.

- Chip bonding on the implant board was a bit risky because testing possibilities by malfunction are reduced and loose connections are hard to track.

- The antenna might easily get out of tune inside the body because of the parasitic effects in the surrounding tissue. Therefore, antenna tuning must take place at the beginning of every recording session in an automatic manner.

- The firmware on the microcontroller had to be updated in order to be conform with the ISO 14155:2011(E) norm for implantable devices. The version of both hardware and software must be implemented in the firmware in order to simplify traceability during and after clinical investigation.

- Power monitoring (received power from the implant), board temperature measurement, and RF signal quality were implemented to improve efficiency of the developed system and to provide the possibility for dynamic adjustment of certain parameter such as primary power supply during operation of the system.

Fig. 6.13 depicts the updated implant board and the packaged implant module. The MyoImp3 was implanted in two sheep for evaluation of its functionality. Hereafter it was implanted in two rhesus macaques to verify functionality of the implant system and to study EMG signals in targeted movements of the primates' hands. The next Chapter presents more details about the internal EMG measurement system and EMG signal characteristics.

The following section summarizes the boundary conditions for the implant system based on the experience collected during the described implantations.

6.3   System Boundary Conditions and Design Issues

An implantable EMG recording-system would be of benefit for better prosthetic control and advanced functionality of prosthesis. However, the design of such a device showed several difficulties that must be overcome. The volume of the implant module which keeps body irritations to a minimum; optimal implant placement for best functionality concerning power and data transmission as well as body comfort; packaging material to protect the body from toxic board materials and the electronic system from malfunction due to body fluids; implant shape to reduce vulnerability to

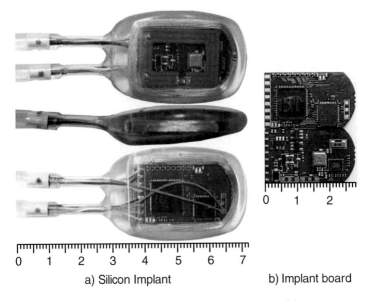

a) Silicon Implant   b) Implant board

Figure 6.13: The MyoImp3 implant electronic module.

mechanical stress from body muscles; implantation method (monopolar, bipolar) for best EMG signal quality and to minimize external noise coupling into the recorded data; reducing thermal effects caused by the energy transmission system; optimized area and power consumption of the implant module; improving system quality in live measurement conditions. To resolve the above mentioned issues, the boundary conditions of the implant system and implantation procedure must be defined and an approach for solving the problems must be conceived.

6.3.1 External Conditions

The external conditions for the implant module mainly concern its geometry and packaging. An area of a 2 euro coin was first specified for the implant module by physicians. The volume was then modified to fit caudal between the scapulae of a primate (rhesus macaque). Therefore, a redesign of the first prototype was necessary as depicted in Fig. 6.14. However, the geometry must avoid sharp angles to reduce the risk of damaging the surrounding tissue and the risk of leakage when

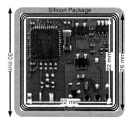
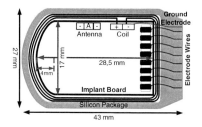

(a) *Design concept of the electronic implant*   (b) *Design concept of the second version of*
in silicone package.   *the electronic implant in silicone package.*

Figure 6.14: Silicone package design of implant prototypes.

using silicone package due to the continuous movement of the neighboring muscles.
Therefore, a round-sided edge was proposed for the board (Fig. 6.14a vs. Fig.
6.14b). The package further reduces the sharpness by rounding the edges. A thickness of $1cm$ is tolerable as well. A big concern from implant with big volume is the
irritation of the animal which might bring the animal to start scratching the skin.
This would lead to wounds and infection risk. The same problem applies to the
electrode wires. Mechanical stability of the electronic board is likely to be affected
by reducing the thickness. Especially when using a silicone package[3].

The implantation period was defined to be short (several months). Therefore,
a silicone package easy to implement was chosen. The interfaces between silicone
and other materials such as plastic (the connector to the electrodes is made of plastic) introduce a weak spot concerning sealing. Fluid might crawl (creep) through
on these interfaces and reach the electronic system causing malfunction. This aspect must be considered especially when using multiple parts for implantation such
as: Implant module and electrodes connected through a connector to the implant
module. Sealing of the connector part takes place during the surgical operation.
The proof of hermetic seal is difficult, though. Furthermore, the electrode wires
to connect the implant to the connector have one isolation layer which possesses
interfacing between silicone and isolation material. A pump effect through the continuous muscle movement in the surrounding area might pump body fluids to the
implant electronic causing malfunction. Using an implanted system as one entity

---

[3]A metallic package such as Kovar, titan, or ceramic introduce higher mechanical stability and
long term implantation possibility due to high hermetic sealing properties.

(one part) would be the best way to ensure hermetic sealing. Thus, it would aggravate the surgical procedure and complicate the tunneling from electrode position to the implant module position. By partial damage during surgery the whole implant system must be exchanged.

The implant placement is dependent on the volume dimensions of the packaged module. Furthermore, the closest spot to the target muscles was identified according to the body dimensions of the animal. One rhesus macaque was rather thin. Therefore, the electrode wires were tunneled to the muscle below the right scapula. The implant was inserted and fixated under the muscle close to the surface. Another rhesus macaque had more volume between the scapulae. Therefore, the wires were tunneled to the implant module that was placed caudal between the scapulae. The decision on placement was sometimes done during the surgical operation. However, the electrode wires were lengthened to have extra space for displacement.

Implant placement affects some electrical parameters in a wireless application. The efficiency of the RF-antenna depends on its radiation characteristics. The implant antenna was placed close to the skin for higher RF-signal strength and hence reliable data transmission.

6.3.2 Internal Conditions

The internal conditions concern the effects on the electronic system which might have an impact on the functionality of the implant. Furthermore, it contains the electronic system characteristics which might affect the biologic surroundings such as heat through eddy currents from the energy transmission system (internal effects from the electronic system on the biological surroundings).

Energy Transmission System

A wireless supplied system needs to be carefully designed. Several issues must be considered during conception such as heat [104] [105] [106] [107] [108]. An electromagnetic field (emf) is induced in the implant board copper planes due to a sinusoidal current in the primary coil following Faraday's law. As a result eddy current flows in the planes as depicted in Fig. 6.15. Eddy current leads to a power loss in the implant copper plane due to a small resistance of the current path in the plane, $R_{eddypath}$ and hence would heat up the implant board. By the addition

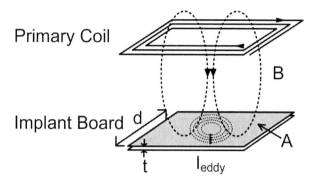

Figure 6.15: Eddy current coupling.

of all eddy currents in the planes we can calculate the overall power loss and hence the heat on the implant board caused by the primary coil. Heat caused by high losses might affect the body tissue in the immediate vicinity. Therefore, the amount of heat must be reduced to allowable levels in order to prevent any damage to the surrounding tissue.

The losses are mainly dependent on the magnetic field produced by the primary coil and the frequency of its current. The loss in the implant copper plane is described in [109] as:

$$W_e \propto f^2 B^2 d^2 t \tag{6.4}$$

where f is the frequency, B is the magnetic flux density, d is the width of the metal plane, and t is the thickness of the metal plane. Power losses due to eddy current are dependent upon the *square* of the frequency, flux density, and the geometry of the metal plane (width, thickness). The usage of 13.56 MHz frequency would increase the losses 11768 times compared to a frequency of 125 kHz when keeping all other parameters the same.

Based on the distance and angle conditions[4] of the energy transmission system, a strong magnetic flux density has been used. The choice of the low frequency of 125 kHz was done to reduce power losses due to eddy current. Furthermore, the local temperature on board has been read using the MSP430 microcontroller to

---

[4]The distance between the external and internal energy coils changes upto 8 cm and the angle changes upto 45° in the measurement setup for animal experiments. The final product would have considerably less distance and a stable angle which allows for more efficiency in power transmission.

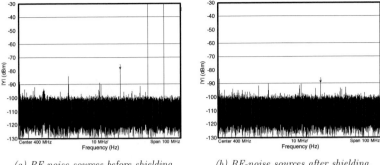

(a) RF-noise sources before shielding.   *(b) RF-noise sources after shielding.*

Figure 6.16: RF-noise before and after shielding.

monitor the heat on board caused by eddy current.

Another effect of the energy transmission system was due to the coupled voltage on the electrode wires placed in the near field of the magnetic field of the primary coil. A big coupled voltage might saturate the input amplifiers and hence aggravate the whole recording. Therefore, a passive filter was implemented at the input of the analogue channels.

Two possibilities have been introduced to design the energy coil of the implant:

1. The coil surrounds the implant board, or

2. Several boards for energy and data antenna are stacked above the implant.

The first method was used in order to reduce the overall thickness of the implant. The PCB antennae would double the thickness of the implant and might invoke irritations for the animal.

High frequency noise sources

High frequency noise sources have been noticed in the measurement room. Several devices use the RF-frequency close to the operating frequency of the RF-transceiver ($402\ MHz$). Fig. 6.16a depicts the surrounding noise frequencies and the strength of these signals in the measurement room. Shielding of the measurement room using aluminum foil has been utilized during the measurement of prototype II of the implant system to reject the RF-noise signals of the electronic devices in the

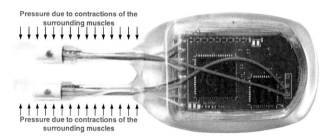

Figure 6.17: Pressure exerted on the electrode wires near the connector.

surrounding area from disrupting RF data transmission. The noise signal strength has been suppressed below $-90dBm$ after shielding (Fig. 6.16b).

In prototype 3 (MyoImp3) RF Saw-filter has been used to filter out RF-noise so further shielding was not needed anymore.

RF Link Quality

The effective data rate of the transceiver depends on the receiver sensitivity. Higher RF signals secure highest possible data rate. Due to absorption and reflections between the body tissue layers the RF signal loses strength. As a result the bandwidth of the data link is reduced. Moreover, mismatch and parasitic effects from the dynamical surroundings of the antenna (ionic body fluids) influence the RF-antenna and might bring it out of tune.

In order to assimilate the data rate of the recorded internal EMG to the wireless available bandwidth, the firmware on microcontroller provide the possibility to change the data rate by down-sampling and reducing the bit resolution of the recorded EMG (see chapter 7.1.2).

Mechanical Stability

The soldered wires on the implant board which connect the electrodes experience a continuous pressure from the body muscles. It can be expressed as a bending of the two flexible parts to the connector as depicted in Fig. 6.17. A break or damage at the soldered or welded locations (on implant board or at the connector) is therefore possible. To relieve the stress the wires had to be routed with a small loop then soldered on the appropriate pin on board (strain-relief).

The thickness of the electronic system board (PCB) is critical. Due to mechanical stress from the body muscles, silicone-packaged[5] components on board (such as capacitors, resistors, or inductance) may be damaged (Fig. 6.18). Increasing the thickness provides more mechanical stability and works against bending the board through external muscle forces (MyoImp2 vs MyoImp3: thickness 0.5 mm vs. 1 mm). Thus the overall implant volume and the need for small electronic circuit components increase.

Figure 6.18: Broken capacitance on a thin implant electronic board.

Artifacts

The strong contractions of the surrounding muscles usually lead to rash movement of the electrode wires and the connector relative to the implant recording board. As a result a change of coupling properties on the electrode wires occurs. Therefore, movement artifacts appear in the shape of a spike at the differential recorded signal. Strong movements might even saturate the analogue amplifier of the conditioning circuit as depicted in Fig. 6.19. Several works have suggested to remove motion artifacts from bio-recordings [110] [111] [112]. Reducing the effect of motion artifact is best implemented by considering the mechanical stress on the wires in advance. The route of the wires from the electrode location to the implant electronic module must be carefully chosen to reduce pressure effects from body and muscle movements. Furthermore, the usage of connectors introduces more vulnerability to the system depicted in slack joint (loose contact). A smooth change in silicone thickness is advisable to reduce the pressure influence on a certain location.

---

[5]A metallic package would mechanically protect the implant board from damage.

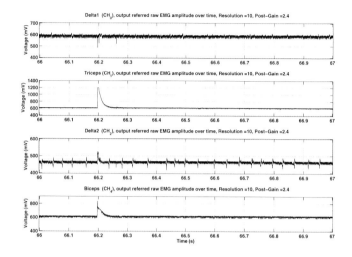

Figure 6.19: Movement artifact from a primate recording.

During the implantation in a rhesus macaque increasing emergence of spike-artifacts contaminated the recorded signals in a large manner which made the signals useless. An artifact investigation showed a remarkable increase of artifacts in the recorded signals whenever the connector or the implant position is externally pressed. However, the external pressing on the electrode showed a smaller number of artifacts. Thus, the connector seems to exhibit a weakness in the designed system which might affect the quality of the recording due to movement.

6.4   Implant System Evaluation

The implant system has been used in several animal experiments for targeted EMG signal acquisition. An implantation in two sheep (data was recorded for more than three months) and two interventions into primates (data was recorded for a total of more than one and a half year). The results show good consistency and long endurance despite the fact that the system was silicone packaged. It delivered a lot of EMG data under noisy environment. However, the usability of the system needed some training for the staff in the recording center.

A few groups world wide follow the idea of controlling a prosthesis with the use of an implantable solution with reasonable concepts and mature results for a medical product. The "Implantable MyoElectric Sensor" (IMES) system of Weir et al. [35] acquires internal EMG signal from up to 32 positions. It utilizes a sensor module (length x diameter: 15 mm x 2 mm) equipped with two electrode contacts for recording EMG activity and an ASIC for conditioning the acquired activity. A big external coil is used on the stump of the arm for wireless power transmission. The power field via the power coil is modulated to send control signals to the implanted modules. Two bands can be used for wireless data transmission: band 1 (60 kHz) with low data rate of upto 6.144 *kbps*, and band 2 (6.8 MHz) with higher data rate of upto 120 *kbps*. The big external coil is associated with efficiency issues. It must provide the IMES sensors in the implanted region with power. However, the implanted region could be big depending on the existent amputation level. Furthermore, the IMES sensors must be placed parallel to the external coil for optimal power transmission. This cannot always be guaranteed due to muscle shape and some external effects in daily life. Therefore, the power transmission efficiency of the IMES system is lower than a centralized power consumption with a specified minimum distance between internal and external coils. The lack of an anti-aliasing filter in the IMES system might introduce aliasing when increasing the number of sensors as the sampling frequency per channel would increase due to the restricted data rate.

The IMES system was probed in cats by Weir et. al. [35]. Furthermore, Baker et. al. [36] implanted the IMES system in a rhesus macaque for more than two years. The movement of the thumb, the index and the middle finger was detected and decoded using an offline parallel linear discriminant analysis (LDA). The drawbacks of the external power coil were stated in the work of Baker et. al. [36] as one IMES sensor was located outside the magnetic field of the external coil and another sensor didn't send data continuously due to the lack of sufficient power over time. McDonnall et. al. [113] developed a centralized implant system and tested it in six dogs for a period of one week in each dog. The system is made of a printed coil and an electronic implant system side by side on the same substrate. As a result the area and the volume of the implant system is enlarged which restricts the choice of the location for implantation. Commercial parts have mainly been used in the system which aggravates the miniaturization of the system. Furthermore, the stability of

Table 6.2: Implant system comparison.

	[35]	[36]	[113]	MyoPlant	comment
Channel number	32	8	4	4	
Bit resolution	8	8	12	10	
Data rate	120 kbps	81 kbps	96* kbps	143 kbps	*: ADC sample rate $2kS/s$
Bandwidth	4 ... $70Hz$ - $200...6.6kHz$	4 ... $70Hz$ - $200...6.6kHz$	20 - $300Hz$*	$7 - 800Hz$, $7 - 1.5kHz$	*: estimated
Gain	$19 - 78dB*$	$19 - 78dB*$	46dB	$33 - 62dB*$	*: tunable
Noise	15 μV_{rms}	15 μV_{rms}	2.2 μV_{rms}	2.1 $-$ 2.4 μV_{rms}	
CMRR	-	-	$> 55dB$	$> 69dB$	
Power	-	-	$300mW*$	Chip: $1.8mW$, Implant: $30mW$	*: external
Dimension	15x2 (LxD)	15x2 (LxD)	70x35x? (LxWxH)	40x27x10 (LxWxH)	L: length, W: width, H: height, D: depth, in [mm]
Anti-aliasing filter	-	-	3rd order, $60dB/dec$	5th order, $100dB/dec$	

the system for a long period of time was not tested due to the short implantation time.

Table 6.2 summarizes the most important characteristics of the systems mentioned above and of the MyoPlant system. The MyoPlant system exhibits better properties concerning data rate, noise, power consumption, and filter order.

In-vivo Implantation

The electronic implant-system was probed in two animal models and four implantations. All implanted systems were hermetically sealed in a silicone package. The implantation experiments are listed as follows in chronological order:

1. **First implantation (Rhesus macaque):** The first implantation in a rhesus macaque delivered a small amount of data. Due to problems in data and power transmission it was not possible to record data in a stable manner two weeks after operation.

2. **Second implantation (Rhesus macaque):** The first implanted system was removed and a modified version of the first system has been successfully implanted in a rhesus macaque. Data has been gathered for about six months at regular intervals. After six months no contact could be made to the implant system anymore. A short circuit on the implant board had caused a malfunction. Body fluids were noticed on the implant board as depicted in Fig. 6.20. The implant board worked fine after removing the body fluids.

3. **Third implantation (sheep):** The functionality of the implant system has been tested in two sheep. Data has been gathered successfully over three months.

4. **Fourth implantation (Rhesus macaque):** The malfunctioning implant of the second implantation has been replaced by a revised version of the implant module. Another implant has been inserted into a second rhesus macaque as well. Data has been successfully recorded for more than six months from both animals.

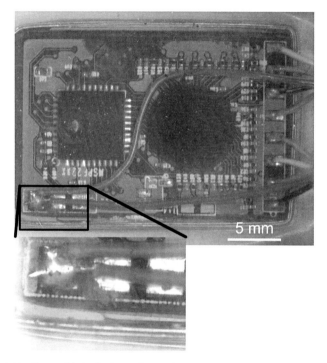

Figure 6.20: Trace of body fluids on the ex-planted implant board which caused a short circuit.

Chapter 7

Digital Control Unit

The implant system utilizes a microcontroller to collect the digitized bio-signals from the MyoC1-chip and send them to the wireless RF-chip (Zarlink transceiver [70]) for wireless transmission to an external receiver system (Fig. 3.1). In order to achieve this a Texas Instruments low-power microprocessor from the MSP430 family is used. It implements a protocol for the communication with the external system connected to a PC (base station). Thus it is possible to externally configure the conditioning circuit for optimized operation and control the output data rate of the system to assimilate it to the available bandwidth of the RF-transceiver. The microcontroller consumes low power and fulfills the time constraints for acquiring the data from the ASIC and forwarding them to the RF-chip using a clock frequency upto 16 MHz.

7.1 Communication Protocol

EMG data transfer between the implant and the external application employs byte streaming in order to attain a maximum utilization of the available bandwidth for EMG data. The external application may send a configuration packet "C-packet" for setting up the implant's operating mode. The implant answers every configuration packet with an acknowledgement called "I-packet" to inform the external application of its actual configuration. According to the commands in the "configuration packet", the microcontroller sets up the implant data blocks and streams them to the external receiver through the RF-transceiver. Streaming the EMG data depends mostly on the activated analogue channels and the bit resolution of the ADC. Fig. 7.1 shows the available structures of the streamed data packets for a certain number of active channels and a configured bit resolution.

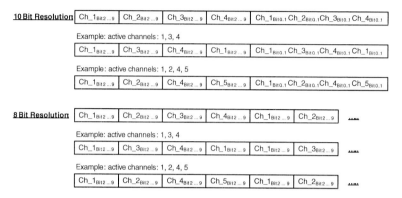

Figure 7.1: Structure of the data packets in streaming mode.

Fig. 7.2 describes the communication setup being initiated by the external application (base station) for acquiring EMG data from the implant. A C-Packet initializes the implant with the desired configuration for EMG data acquisition. The implant acknowledges its received configuration data through an I-Packet. Hereafter, it sends EMG data recorded by the ASIC (MyoC1) according to the setup configuration in a streaming mode.

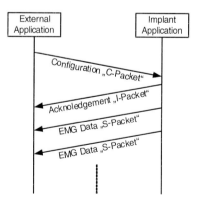

Figure 7.2: Implant configuration for EMG transmission in streaming mode.

7.1.1 Firmware Architecture

The state machine of the microcontroller firmware is depicted in Fig. 7.3. Whenever the implant is supplied with enough power, it initializes the microcontroller ports and the communication interfaces to the neighboring chips on the implant board. The RF-chip is initialized as well in this step by assigning a specific identity number for the RF-chip[1] and configuring the RF-parameter. Before we initialize the wireless communication, the RF-antenna is tuned by a routine in order to choose the strongest RF-signal by matching the impedance of the external circuit elements (including antenna matching network) to the input impedance of the receiver amplifier or to the output impedance of the transmitting amplifier. An internal capacitor bank on the RF-chip is used for the impedance matching. Furthermore, the routine uses an internal ADC on the RF-chip for signal strength measurement. Hereafter, the RF-communication with the external system (base station) is initialized according to the RF-chip manual. The initialization phase is finished with initializing the configuration register of the ASIC with the default parameter for the channel bandwidth, input selection of channel 4 (Fig. 4.31), post-amplifier gain, and the power-down setting for the analogue channels. The firmware then starts a perpetual loop which mainly executes the received command from the external application. The RF communication is continuously checked during command execution through analyzing the interrupts from the RF-chip. The RF-communication is reinitialized when lost.

A 256 byte array for EMG data is managed on the microcontroller to buffer the EMG data from the ASIC in order to avoid data losses in the dynamical wireless connection. When the array is overloaded (full) due to non-optimal wireless conditions, the internal buffer is cleared and the number of lost data packets is saved on the microcontroller. The number of lost packets is then transmitted to the external application per acknowledgement packet (I-Packet) in order to reestablish the time axes of the EMG curve and to be informed about the RF link quality.

The commands from the external application stop the functionality on microcontroller and have high priority. Every command is acknowledged by one "I-Packet" (see Fig. 7.2). Every acknowledgement contains the actual configuration parameters

---

[1]The identification number of the RF-chip (IMD-ID) is identical to the ID number of the MyoC1 chip. Thus it is possible later on to allocate the analogue measurement parameter for EMG signal processing such as channel gain, channel noise, and channel bandwidth.

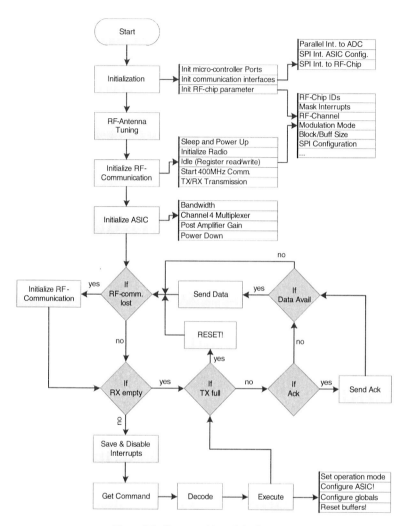

Figure 7.3: State machine of the firmware.

of the implant as explained later on in this chapter. The acknowledgement is sent prior to streaming the EMG data and when overloading the internal data buffer.

The microcontroller sends data through SPI interface to the RF-chip. The data is divided into blocks with a constant length defined at the initialization phase.

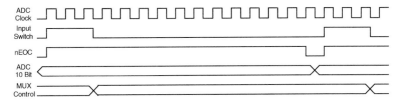

Figure 7.4: Timing diagram for the multiplexer.

However, the RF-chip has its own packet length for RF-communication.

The ASIC digital functionality is controlled mainly by the microcontroller. Especially the multiplexer on ASIC underlies strict timing conditions for optimal operation of the analogue part of the system. The ADC requires 14 clock cycles for one digital conversion. The first 2.5 cycles are used to acquire the analogue value from the post-amplifier. The next channel in the multiplexed configuration of the system must be selected immediately after acquiring the analogue value by the ADC (2.5 clock cycles after conversion begin). The multiplexer control is implemented using a timer on the microcontroller. The timer gives the analogue channel the maximum available time to prepare the next analogue value for digital conversion. This is very important for reliable data acquisition especially at the post-amplifier (see section 4.2.2.5). Fig. 7.4 shows a timing diagram of the multiplexer based on the ADC clock frequency and the end of conversion signal (nEOC) out of the ADC.

The clock of the ADC on the ASIC is delivered from the RF-chip. The microcontroller selects the value of that clock by configuring the programmable output register "PO3" on the RF-chip. Table 7.1 contains the possible clock values and the resulting sample rate at the output of the ADC.

The maximum data rate is restricted through the RF-chip. It is possible to transmit a maximum of 250 kbps. To achieve this a default clock configuration of 200 kHz is used during live measurements.

7.1.2   Protocol Operation

Whenever the external application sends a configuration packet to the implant system, the RF-chip releases an interrupt to the microcontroller signaling that data must be read. The microcontroller reads the coded data from the RF-chip and decodes it. The configuration packet from the external application consists of a block

Table 7.1: ADC clock configuration and the associated data rate.

Clock	Data Rate
150 kHz	10.7 kS/s
200 kHz	14.3 kS/s
300 kHz	21.4 kS/s
400 kHz	28.6 kS/s
600 kHz	42.9 kS/s
800 kHz	57.1 kS/s
1.2 MHz	85.7 kS/s
2.4 MHz	171.4 kS/s
3 MHz	214.2 kS/s
6 MHz	428.6 kS/s
12 MHz	857.1 kS/s

of 15 bytes of coded data. The data is coded as follows:

- **Byte 1:** Not used.

- **Byte 2:** The command code is given in byte 2. It is coded as follows: A brief

Table 7.2: Command code.

Command	Code [HEX]
Reset	0x01
Configure Implant	0x08
Stop	0x10
Start	0x20
Test Mode	0x30
Energy Monitoring	0x40
Implant Status	0x50

explanation of used commands and their semantic is explained later on.

- **Byte 3-7:** It concludes a 40 bit value (5 bytes) of the ASIC configuration register (see section 4.2.4). It consists of the configuration parameter for the channel bandwidth (7 - 800 Hz or 7 - 1500 Hz), channel 4 selector, the gain of the post-amplifier (1X, 2.4X, 12X, 24X), and the power-down setting of the analogue channel block.

- **Byte 8:** Active channels on ASIC and the bit resolution. The first five LSBs contain the channel state (active = 1, not active = 0. See Table 7.3) and the three MSBs contain the bit resolution code as depicted in Table 7.4.

Table 7.3: Active Channels and bit resolution.

Bit number	Functionality
0	Channel 0
1	Channel 1
2	Channel 2
3	Channel 3
4	Channel 4

Table 7.4: Bit resolution code.

Bit[7],Bit[6],Bit[5]	Resolution
000	Channel 8
001	Channel 10

- **Byte 9 (Index):** The number of measured values per cycle for the coupled voltage on the implant. The coupled energy monitoring (or voltage monitoring) on the implant is measured x-times and then it is transmitted to the external application during the energy monitoring mode. The number of measured voltages per cycle is determined by byte 9.

- **Byte 10:** Data rate for the test mode. The test mode is used to determine the maximum possible data rate without losses. The microcontroller generates 8 bit values (up counter) and sends the generated data to the RF-chip. The data rate of the generated data is set using byte 10. Table 7.5 contains the codes for the available data rate configurations.

- **Byte 11:** Multiplexer timer value. This byte contains the time value for the 2.5 cycles needed from acquiring the 10 bit ADC output till switching to the

Table 7.5: Data rate configuration for the test mode.

Byte 10	Data Rate [kbps]	Byte 10	Data Rate [kbps]
$0x00$	10	$0x0D$	130
$0x01$	10	$0x0E$	140
$0x02$	20	$0x0F$	150
$0x03$	30	$0x10$	160
$0x04$	40	$0x11$	170
$0x05$	50	$0x12$	180
$0x06$	60	$0x13$	190
$0x07$	70	$0x14$	200
$0x08$	80	$0x15$	210
$0x09$	90	$0x16$	220
$0x0A$	100	$0x17$	230
$0x0B$	110	$0x18$	240
$0x0C$	120	$0x19$	250

next analogue channel at the multiplexer by the microcontroller (see Fig. 7.4). The timer value is dependent on the used ADC clock.

- **Byte 12:** Analogue channel mask on the microcontroller. The activated channels in byte 8 can be masked in byte 12 (0 not active on microcontroller, 1 active.). The main advantage of masking the acquired data from the analogue channels on the microcontroller is to reduce the overall data rate in case the RF-bandwidth is reduced due to surrounding effects such as noise or antenna orientation.

- **Byte 13:** Down sampling. Another way to reduce the effective data rate for the RF-link is to reduce the sampling frequency on the microcontroller by not saving every sampled value by the ADC. Byte 13 describes the sampling ratio $(1/2, 1/3, 1/4 \cdots , 1/10)$ on the microcontroller.

- **Byte 14-15:** Not used.

The operation modes of the firmware are given as follows:

- **Acquire EMG Data:** The command "Configure Implant" sets the microcontroller into EMG acquisition mode using the configuration parameter in the configuration packet. The configuration parameter mainly sets the ASIC into the desired operation mode. Besides, it is important on the external application to select the active channels on ASIC and on microcontroller and the desired bit resolution. Reducing the sampling rate can be used by byte 12 and byte 13 in case the RF-bandwidth is not optimal. The implant writes the received ASIC configuration into the digital control register of the MyoC1-chip and initializes its EMG data buffer. Furthermore, the interrupts and the timer parameter are reinitialized according to the received configuration. EMG data acquisition is started and an acknowledgement block is sent prior to streaming the EMG data.

- **Test Mode:** The test mode is used to determine the maximum possible data rate of the RF-link without losses. The microcontroller generates 8 bit values (up counter) and sends the generated data to the RF-chip for wireless transmission. The data rate of the generated data is set using byte 10. Table 7.5 contains the codes for the available data rate configurations. The test mode is used prior to live measurement to verify the maximum bandwidth of the RF-link with no losses.

- **Energy Monitoring:** The coupled energy on the implants coil varies due to distance and angle variation between the external and the internal coil (see section 5.1). The ADC on the microcontroller is used to measure the coupled voltage and hence the coupled energy on the implant. A voltage divider (4:1 ratio) is used before the voltage regulator and the divided voltage V_{div} is connected to the input of the ADC on the microcontroller (see Fig. 7.5). The 10 bit digital value of the divided voltage is then transmitted to the external application to verify the energy transmission system. The coupled voltage at the input of the voltage regulator V_{coupl} is calculated from the digital value of the divided voltage V_{div} as follows:

$$V_{coupl} = 5N_{div}\frac{V_{DD}}{1023} \tag{7.1}$$

where N_{div} is the measured 10 bit digital value of the divided voltage V_{div}, $R_1 = 4R_2$ in Fig. 7.5, and $V_{DD} = 3.3$ V is the supply voltage of the ADC on the microcontroller.

- **Implant Status:** The microcontroller continuously sends an "I-Packet" inside

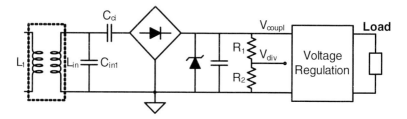

Figure 7.5: Schematic of the energy receiver system including the energy monitoring voltage divider.

the implant status mode. This includes the ASIC configuration, the coupled voltage from the energy transmission system, RF link quality information (register values in the RF-chip concerning link quality), and the temperature of the implant board measured locally by the microcontroller.

Table 7.6 summarizes the packet types that can be sent from the external application to the implant system.

The external application must configure the implant so that a loss-free transmission of all data under the given measurement conditions is possible. It monitors the achievable data rate of the RF-link compared to the source data rate out of the ASIC. In case both values are similar the source data rate can be decreased by reducing the bit resolution or reducing the sampling rate (down-sampling on microcontroller).

The internal EMG data buffer on the microcontroller is cleared whenever it experiences an overflow caused by bad RF link quality. The microcontroller signals to the external application an overflow whenever possible by sending an "I-Packet" containing the number of resetting the internal buffer.

The acknowledgement packet "I-Packet" is sent by the implant to the external system after the start-up, a reset command, a configuration command (Config., Test, Energy, and Status Mode), and after overloading the internal buffer on the microcontroller. The "I-Packet" is created as follows:

- **Byte 0:** The first byte in the 15 bytes block has the following value:

  1. $0x00$: indicates a data packet.

  2. $0x01$: indicates an "I-Packet"

Table 7.6: Types of the packets from the external application "C-Packet".

Byte							
0	not used	not used	not used	not used	not used	not used	not used
1	Reset	Config.	Stop	Start	Test Mode	Energy Mode	Status Mode
2	ASIC configuration: 40 bit						
3							
4							
5							
6							
7	Active channels and bit resolution						
8	Index						
9	Data rate						
10	Multiplexer timer						
11	Channel mask on microcontroller						
12	Down-sampling on microcontroller						
13	not used						
14	not used						

- **Byte 1:** The version of the firmware and that of the hardware are stored in this byte. The four MSBs contain the hardware version and the four LSBs contain the firmware version. The utilization of a version number is part of the ISO 14155:2011(E) "A.2 Identification and description of the investigational device" for medical devices. Hereby it is possible to track the hardware and the firmware of the implanted device to permit a full identification.

- **Byte 2:** The reset counter includes information about how often the internal EMG data buffer has been overloaded and then cleared by the firmware. The external application is then capable of calculating the amount of data lost due to imperfect RF link.

- **Byte 3-6:** The ASIC configuration without the last unused byte is transmitted back to the external application.

- **Byte 7:** Active channels on ASIC and the bit resolution are stored exactly as in byte 8 of the "C-Packet" from the external application. In the "Status Mode" this byte contains the MSB byte of the coupled voltage measured by the ADC on the microcontroller.

- **Byte 8:** The LSB byte of the coupled voltage measured by the ADC on the microcontroller is placed in byte 8. The digital converted value of the coupled voltage is 10 bit.

- **Byte 9:** It contains one of the following parameters:

    1. Link Quality: The sum of interrupts concerning bad link quality.

    2. Data Rate: Under "Test Mode" it contains the data rate setting.

    3. Index: Under "Energy Mode" it contains the number of measured values per cycle for the coupled voltage on the implant.

- **Byte 10:** It contains one of the following parameters:

    1. Multiplexer timer value.

    2. Maximum bit error counter under "Status Mode" which contains the sum of link quality interrupts triggered by the RF chip after crossing the maximum block error in one package transmission.

- **Byte 11:** It contains one of the following parameter:

    1. Channel mask on the microcontroller.

    2. Under "Status Mode" it contains the sum of link quality interrupts triggered by the RF chip after crossing the maximum number of packet retransmissions.

- **Byte 12:** The four MSB bits contain the bit[9] and bit[8] of the coupled voltage on the implant sampled by the ADC on the microcontroller. The four LSBs contain the setting of the down sampling byte.

- **Byte 13-14:** The locally measured temperature (10 bit) by the microcontroller.

Table 7.7 summarizes the packet types that can be sent from the the implant system to the external application.

Table 7.7: Types of the acknowledgement packets from the implant firmware "I-Packet".

Byte	Reset	Config	Stop	Start	Test Mode	Energy Mode	Status Mode
0	0x01						
1	Version						
2	Reset counter						
3	ASIC configuration 40 Bit						
4							
5							
6							
7	Active channels / Bit resolution						Energy MSB
8	Energy LSB						
9	Link quality				Data Rate	Index	Link quality
10	Multiplexer timer						Max. berr counter
11	Channel mask						Max. retries counter
12	Energy MSB / Down sampling						
13	Temperature MSB						
13	Temperature LSB						

Chapter 8

Experimental Results and Signal Properties

Losing an upper extremity is a crucial incident causing huge restrictions for casualties in daily life. The loss of an upper extremity either due to accidents, war or to amputation caused by trauma, disease, tumor or it is congenital [9]. Patients are in most cases provided with a state of the art artificial hand prosthetic. The measurement of muscle activity remains the source of controlling most of the present-day state of the art hand prosthetic systems [10]. A surface electromyogram (sEMG) is often used by placing surface electrodes on the skin overlying the target muscle. The control mechanism of hand prosthetic employs independent source signals to achieve certain degrees of freedom (DoF) in the motion of the prosthetic hand. Generally, two independent signals (or rather independent signal components) are needed to accomplish one DoF. The measured signal generally contains accumulated activity of many motor units in the underlying recording region and less from deep muscle regions. Besides, the high frequency content of surface-recorded biopotentials is smoothed according to the low-pass filter characteristic of several skin layers. Therefore most of the spectral power of the sEMG is located below 400-500Hz which leads to a sampling rate of approximately 1 kHz [11], [12]. Cross talk and interference from neighboring muscles add up greatly to the surface-recorded signal caused by the relatively big recording area covered by the surface electrode. Moreover, interference from several muscle groups reduces the selectivity of sEMG. Long-term sEMG recording leads to skin irritations caused by surface electrodes. Comfort of hand prosthesis is therefore affected. Artifacts are a major concern in sEMG recording. Movement artifacts initiated by electrode movement relative to the skin or cable movement [13], electromagnetic coupling (50/60 Hz), and variable impedance of the skin interface caused by sweat are some of the issues affecting long-term signal quality of an external EMG.

The MyoPlant project proposes a centralized implantable approach to overcome the difficulties of an external surface recording as explained before and hence improve recorded signal quality. Invasive EMG is less vulnerable to electrode movement artifacts and offers an opportunity for recording from deeper muscle regions [14]. Furthermore, measuring independent muscle control signals is essential for a prosthetic hand with multiple degrees of Freedom. An invasive recording would improve the quality of the recorded signals and possibly offer a highly selective EMG for advanced and smooth control of a hand prosthetic. Improvement of signal stability and robustness over time for long-term recordings due to stable electrode-tissue interface is expected.

The implantable system is divided into several parts or modules. The first module is the electrode interface followed by an electronic implant system for biopotential signal conditioning. An external module is used for wireless power and data transmission. The electrode interface utilizes epimysial silicon electrodes to acquire localized muscle activity. The electronic system contains an integrated circuit (IC) for signal conditioning (amplifying, filtering, and digitizing biopotential signals), an energy management circuit, a wireless transceiver for data transmission, a digital controller to coordinate data transmission on the electronic system and to control the communication with an external application. An external module is used to transmit enough power to the implant electronic system under dynamic movement conditions where the distance and angle between external and internal coils varies due to animal movement. Furthermore, it includes a wireless system for bidirectional signal transmission between the implant electronic module and the external application running on a PC station. The recorded signals are analyzed to verify signal content and quality.

8.1   System Evaluation

An animal experiment on a rhesus macaque was firstly done to test the initial electrode design based on a flexible polyimide structure described in [33] and [5].

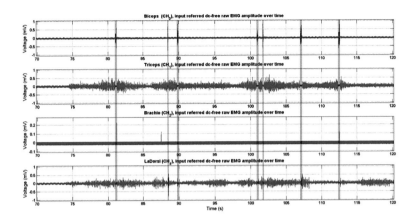

Figure 8.1: *Internal EMG from a sheep.*

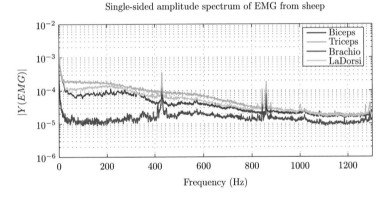

Figure 8.2: *Internal spectrum of EMG from a sheep.*

As mentioned in chapter 3.1, the bonding of the wire to the stripe electrode structure was vulnerable to strong muscle movement. This led to a break in the bonding plate and to the loss of contact with the electrode sites. To overcome the mechanical stability issues with the flexible polyimide-based micro-electrode an epimysial silicon electrode was designed and evaluated in [114]. The epimysial silicon electrodes were implanted in 14 rats for a period of 8 weeks to validate functionality and evaluate the electrodes electrically and mechanically after implantation and to check

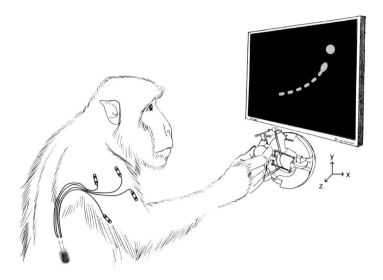

Figure 8.3: The first rhesus macaque in the measurement setup.

a developed surgical procedure [67]. The implant electronic system including the epimysial electrodes were then implanted in two sheep to validate the functionality of the whole implantable system including silicon electrodes and implant electronics. The implantation procedure was successfully repeated in the sheep implantation. Electrode impedance measurement was utilized during the surgical operation to verify the functionality of the implanted electrodes. EMG measurement results from the sheep proved the functionality of the system and EMG signals were recorded with an amplitude range of $\approx 2~mV_{pp}$. Fig. 8.1 shows the EMG of the sheep during movement (forward stepping). It is possible to recognize the movement of the sheep in the recorded EMG (the red blocks in Fig. 8.1). The spectrum of the recorded signals showed high activity in the frequency range around 200 Hz as depicted in Fig. 8.2. Noise sources from the surrounding devices in the measurement room are seen in the spectrum as peaks in higher frequency range (around 450 Hz, 850 Hz, 1000 Hz, 1150 Hz, 1350 Hz) as well.

As rhesus macaques show big similarities to human beings, the implant system was then implanted in two rhesus macaques using silicon epimysial electrode pairs on the following muscles (Fig. 8.3): a) biceps brachii b) triceps surae c) deltoid

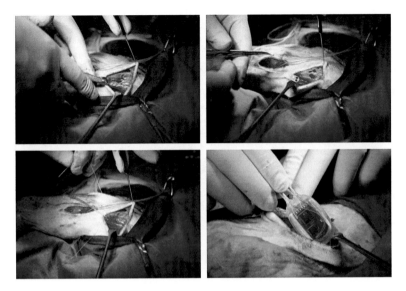

Figure 8.4: Implantation procedure.

clavicularis d) deltoid acromialis. The goal of the animal implantation was to verify the functionality of the implant system for recording internal EMG and to analyze the quality of the recording and that of the collected EMG signals.

The electrodes were inserted longitudinally under the epimysium of the target muscle and fixated using two sutures over the epimysium. Electrode wires were tunneled towards the back. The implant was placed between the scapulae of the first primate and below the right scapula in the second primate because it was rather thinner than the first animal. The implant was inserted and fixated under the muscle by suturing the muscle tissue above the implant. Fig. 8.4 shows the implantation procedure of the EMG implantable system (electrodes and implant placement).

The rhesus macaque is able to execute hand movements similar to that of human beings. The first primate sits in front of a touch screen and touches a lighted circle on the screen during targeted movement as depicted in Fig. 6.11. Digital signals are recorded simultaneously to the EMG recording which are later on synchronized on the PC in order to verify the findings.

A haptic robot is used for the second trained primate to do a cursor task by executing arm movements in 3D directions using the robot's knob as depicted in Fig. 8.3. The robot system captures the position, force, and velocity parameters of the movement among other signals. Furthermore, it is possible to vary the force momentum of the robot's knob which makes it possible to examine the variation in the contraction level of the muscle when moving the knob to the desired location. The animal was trained to move the knob to a target point depicted on a screen in front of the animal. The implantable EMG recording system and the robot acquisition system were synchronized in time using digital signals generated on PC. The time synchronization can then be used for investigating the internal recorded EMG with respect to simultaneously recorded parameter (force, velocity etc...).

The recorded EMG signals from the implant system were post-processed using a digital equiripple bandpass filter between $20 - 1500 \; Hz$ to remove motion artifacts below $20 \; Hz$ and unneeded high frequency components. The root-mean-square (RMS) of the recorded EMG was calculated as follows:

$$\text{RMS} = \sqrt{\frac{1}{N} \sum_{n=1}^{N} s_n^2} \tag{8.1}$$

where N is the number of samples in a time segment ($t_s = 150ms$), s_n is the invasively recorded EMG signal. The result is depicted in Fig. 8.5 for the EMG signal from the biceps brachii muscle along with the RMS of the EMG signal and the measured force in z-direction. A high correlation can be detected between the recorded force and the EMG signal energy (RMS) as depicted in Fig. 8.5 ($Corr(RMS[EMG], Force[z]) > 0.8$). This might help to develop appropriate algorithms for precise control of prosthesis as the activation and deactivation of the muscles are precisely measured. Furthermore, it may offer an easy and reliable way to control a prosthesis with high degrees of freedom using non-correlated multi-channel EMG-recording.

8.2   EMG-Recording from Primates

The implantable system was successfully implanted in two rhesus macaques. No indication for infection or change in signal quality have been detected during and after the implantation. It was possible to record EMG signals from the two rhesus macaques using the implantable system. At the same time other parameters were recorded parallel to the internal EMG from two different measurement systems that

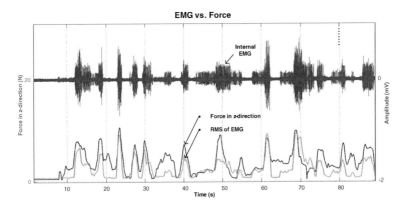

Figure 8.5: EMG from the biceps muscle vs. the force in z-direction and the RMS of the EMG signal.

helped during EMG signal processing and verification phase. A surface EMG recording was initiated simultaneously to the internal recording for comparison purposes. The characteristics of the internal recorded EMG are discussed in this section. Furthermore, the internal recorded EMG is compared to the external one to highlight differences.

Measurement Setup

Internal EMG signals were recorded in bipolar configuration as described in chapter 2.2.1. Bipolar recordings were carried out by using silicon electrodes (see chapter 3.1) with an inter-contact distance of 1 *cm* as depicted in Fig. 3.2. The silicon electrode was inserted under the epimysium of the target muscles (biceps, triceps, deltoid (frontal), and deltoid (back)) as illustrated in Fig. 8.4. The configuration parameter of the implant electronic module were for most of the recording sessions as follows:

- **Number of channels:** Four channels were used to record EMG signals from four different muscles. Channel 2 has a DC-offset which affects its operation when using high amplification factors. This made the acquisition from the deltoid muscle (back) useless.

- **Gain:** The gain of the amplifier circuit was varied between 100 and 600

$(40\ dB - 55.5\ dB)$. $55.5\ dB$ was used in most acquisitions.

- **Bit resolution:** 10 bit resolution was used.

- **Data rate:** The frequency of the ADC was set to $200\ kHz$ which yields a data rate of $14\ kSps$[1] $(= 140\ kbps)$.

- **Signal bandwidth:** The acquisition channels have a bandwidth of $7\ Hz - 1500\ Hz$. A low-pass filter of 5th order is used.

- **PC software:** The wireless collected signals from the implant module were stored on PC in log files with the help of modified software of the wireless transceiver (ZL7010X ADK 2.0.0, Microsemi corporation).

Surface EMG measurement was accomplished with the help of a commercial bio-signal acquisition device (g.USBamp, g.tec medical engineering). The main characteristics of it's recording channels is as follows:

- **Number of channels:** 16 recording channels are available with $\mp 250\ mV$ input signal range, and an input impedance of $10^{10}\ \Omega$. Only four channels were used during the live measurements.

- **Bit resolution:** A separate ADC is used for every recording channel. The ADC has a bit resolution of 24 bit.

- **Sampling frequency:** The surface EMG was sampled at $4.8\ kHz$.

- **Signal bandwidth:** The recorded signals were band-pass filtered between $2\ Hz - 2\ kHz$. A notch filter at $50\ Hz$ has been used to reject the strong $50\ Hz$ coupled power line.

- **PC software:** The EMG recording device was connected to the computer via USB. A software on PC (g.Recorder V2.09a, g.tec medical engineering GmbH) captures the recorded signals and store them in HDF5 file format.

8.2.1   Time and Frequency Illustration

Fig. 8.6 depicts the internal and external parallel recorded EMG from the biceps muscle of the rhesus macaque. Furthermore, it shows the same signals filtered by

---

[1]The ADC needs 14 cycles to convert one 10 bit value into digital form.

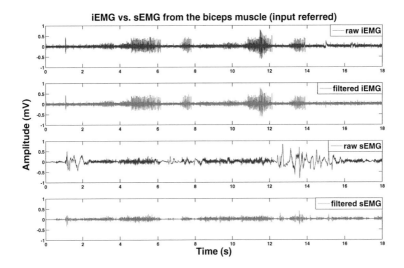

Figure 8.6: Input-referred internal and surface EMG from a rhesus macaque before and after band-pass filtering.

using a digital equiripple bandpass filter between $20 - 1500\ Hz^2$. The input-referred EMG amplitude from the internal measurement (implant system) is twice as big as the surface-recorded EMG amplitude ($\approx 2\ mV_{pp}$ vs. $\approx 1\ mV_{pp}$). Surface EMG shows a higher energy content in the frequency region below 20 Hz as compared to the internal EMG prior to filtering (Fig. 8.7). Signal activity below 20Hz in EMG recording is usually interpreted as a motion artifact. It arises from the movement of the electrodes or the cables during the recording [13] [112]. Motion artifacts are in most cases big enough to drive the amplifier circuit into saturation. Furthermore, it is hard to process surface EMG signals for prosthetic control unless they are filtered. This consumes extra post-processing time and energy. Internal EMG recorded by the MyoImp system provides EMG data ready for processing without the need for extra digital filtering for rejecting motion artifacts due to the fixity of the electrodes and wires to the body tissue and to the high-pass filter of the amplifier circuit. As the internal recording is geometrically close to the origin of activity, it is expected

---

^{2}There is a short delay between the raw signal and its filtered counterpart due to the oscillation delay of the digital filter.

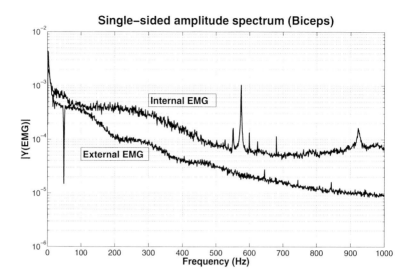

Figure 8.7: Power spectrum of the internal and external EMG from biceps muscle of a rhesus macaque.

to be more reliable with regard to motion artifacts because the wires and the electrodes are more stable and show less movement than their external counterpart due to their attachment to the body tissue.

Nevertheless, motion artifacts exist in internal EMG recording as well due to joint and muscle movement. Therefore, the routing of electrode wires to the central implant module must be well considered prior to implantation. However, the best way to completely reject motion artifacts is by attaching the electronic circuitry directly to the electrode (active electrode). This would increase the wiring overhead between the centralized implant module and the active electrodes. Furthermore, it would make the hermetical sealing of the active electrode more complicated and challenging. Therefore, active filtering of low-frequency noise remains the best solution for now. The MyoC1-ASIC utilizes therefore AC-coupling to reject low-frequency movement artifacts. The high-pass frequency needs though to be increased to approximately $20\,Hz - 30\,Hz$ to completely reject movement artifacts in that range.

8.3 Internal vs. External EMG, Comparison Study

A systematic investigation of the recorded EMG is needed to evaluate the quality of the internal EMG. Therefore, an external EMG recording was carried out parallel to the internal EMG recording in order to better examine the characteristics of the internal EMG compared to state of the art acquisition method, the surface EMG-recording (sEMG). A commercial recording equipment has been used for the external EMG measurement (g.USB, g.tec medical engineering, see section 8.2). The same muscles as the internal recording were externally addressed during the EMG signal acquisition (Biceps, Triceps, Deltoid anterior, and Deltoid posterior).

The compared parameter and the emerging results are explained in the following sections.

8.3.1 Signal Amplitude

The amplitude of the received EMG signal is dependent on the impedance between the source of activity (muscle or nerve cells) and the input impedance of the amplifier (Fig. 1.1). The low-pass characteristic of the skin and the electrode dimension and material are important factors to be considered when measuring the signal amplitude. In addition, the placement of the electrodes on the muscle plays a significant role for the recorded signal amplitude and its frequency components. Bipolar electrodes should be placed 20 mm away from the innervation zone to minimize temporal dispersion for static contractions [115]. This is based on the fact that muscle fiber action potentials (MFAPs) spread bidirectionally from the innervation zone towards the tendon[3] of the muscle. The placement within the innervation zone would lower the amplitude in a differential amplifier. In our experiment the implanted system returned low signal amplitude from the triceps electrode. The reason might be a non-optimal placement of the epimysial electrodes on the muscle. A precise anatomical knowledge of the underlying innervation zones is therefore necessary [116].

Two amplitude values were calculated for the internal and external EMG:

1. **Peak-to-peak voltage (V_{pp}):** The maximum input-referred voltage, V_{pp}, is an important factor for the gain specification of the amplifier circuit. Furthermore, a higher signal amplitude at the input results in higher signal-to-noise

---

[3]The tendon of the muscle is a band of fibrous connective tissue which attaches the muscle to the bone. It's basic function is to move bone due to muscle contraction.

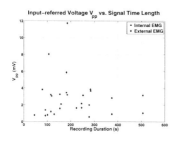 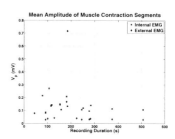

(a) Input-referred peak-to-peak voltage vs recording session time.

(b) Mean input amplitude of the activity Regions (RoA) vs recording session time.

Figure 8.8: Input-referred amplitude of the recorded internal and external EMG from the biceps muscle.

ratio (SNR) with the same amount of noise which in turn leads to better resolution and higher information rate of the recorded activity.

2. **Average amplitude of the muscle activity segments:** The average amplitude of the muscle contraction segments in the recorded signal reflects the mean amplitude difference between internal and external EMG. It is calculated by only considering muscle activity segments of the recorded signal into the average calculation.

Fig. 8.8a depicts the maximum input-referred, peak-to-peak amplitude recorded by the internal and the external EMG-system over the recorded session's time. The internal system shows a stable behavior with low standard deviation of 0.78 mV for the internal recorded muscle activity (iEMG) compared to a standard deviation of 3.3 mV for the surface recorded muscle activity (sEMG). Outliers are frequently observed in the external signal[4] despite the fact that both recorded signals were filtered between 20 Hz − 1500 Hz. The mean value of the peak-to-peak amplitude is calculated for both signals after excluding the outliers. It results in 2.7 mV for the internal EMG versus 1.3 mV for the external EMG. The internal EMG seems to deliver an amplitude twice as big as the external acquired amplitude. The same result is achieved when calculating the average of the mean absolute voltage in the EMG activity regions depicted in Fig. 8.8b (0.13 mV for the iEMG vs 0.057 mV

---

[4]The outliers are high signal amplitudes originated in the change in noise coupling due to movement artifacts of the electrodes or the wires.

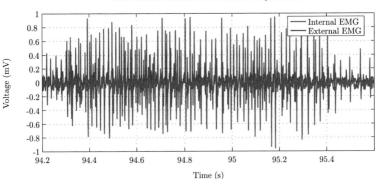

Figure 8.9: Internal and external recorded EMG (amplitude comparison).

for the sEMG). Fig. 8.9 additionally illustrate the amplitude difference between the internal and the external recorded muscle activity.

8.3.2 Signal-to-Noise Ratio (SNR)

The signal-to-noise ratio is an important quantity of a recorded signal. It determines mainly the bit resolution of the analogue-to-digital converter as described in chapter 4.2.3.1. Therefore, the power consumption and the silicon area of the conditioning circuit highly depend on the SNR. Furthermore, the power consumption of the wireless data interface depends also on the SNR of the transmitted data. Reducing the data rate to acceptable limits for instance by coding would reduce the power consumption and relief the power requirements for the implantable device. This leads to higher utilization durability of batteries in upper limb prosthesis.

The SNR is defined as the ratio (logarithmic) of the signal power over the noise power:

$$\text{SNR} = 10 \cdot \log_{10} \left(\frac{P_S}{P_N} \right) \tag{8.2}$$

where P_S is the signal power, and P_N is the noise power. The SNR of the recorded EMG signals (internally and externally) was evaluated after splitting the recorded signal into segments containing muscle activity (contraction of the muscle) and segments containing noise.

The low amplitude level in the external EMG is mostly caused by the high

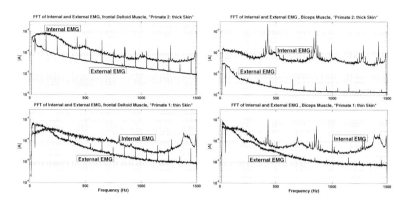

Figure 8.10: FFT of the internal and external EMG from the biceps and the deltoid muscle of two rhesus macaques: a fat (2) and a skinny one (1).

path-impedance between the source of activity and the recording electrodes. Fig. 8.10 shows the FFTs of the internal and external recording in two different rhesus macaques. Primate 2 has a thinner skin than primate 1. As a result the FFT reflects higher activity in the low frequency region in the external recording. Despite the higher amplitude of the internal EMG recordings, the SNR of the external EMG shows a comparable SNR to the internal EMG (Table 8.1). The reason for that is the extra EMG components from neighbored muscles in the surface recording (Fig. 8.11). Besides, a higher noise level in the internal recording further reduces its SNR. This noise is mostly caused by the large coupling from the external energy coil. Furthermore, noise coupling from surrounding devices in the measurement room can be seen in the FFT of the recorded EMGs internally and externally as well.

8.3.3   Selectivity

The animal experiments was initiated to investigate the overall usability of internal EMG for prosthetic control compared to surface EMG. Therefore, different muscles were targeted through implantation. The selectivity of EMG can be tested by implanting several electrodes on the same muscle group, for example three epimysial electrodes on the deltoid muscle (back, middle, front), and verifying the differences between the acquired EMG by means of correlation. Special attention to the electrode size and recording method must be taken when aiming for high selectivity. A

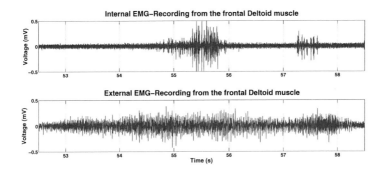

Figure 8.11: A parallel recorded internal and external EMG from the deltoid muscle.

small electrode size and a bipolar recording method improve the selectivity of EMG recording [117], [118]. The MyoPlant system utilizes a bipolar recording method with an electrode area of 7.1 mm^2 (see chapter 3.1).

Despite the high distance between the targeted muscles, it was possible to extract some information about selectivity after executing the surface recording and comparing it to the internal one. Due to the short period of muscle contractions in the internal recording from the deltoid muscle a higher muscle selectivity is believed to exist in the internal recorded signal as compared to the surface EMG (Fig. 8.11). The surface EMG seems to contain a compound of additive muscle activity from higher number of underlying muscle groups which is deduced in the long period of time of the muscle contraction per activation as seen in Fig. 8.11.

Table 8.1: SNR from a rhesus macaque.

Primate	Recording	muscle	SNR	Comment
1	internal	Biceps	13.9 dB	
1	external	Biceps	13.4 dB	
1	internal	Deltoid	15.9 dB	
1	external	Deltoid	13.5 dB	
1	internal	Triceps	10.1 dB	Low amplitude, non-opt. placement
1	external	Triceps	14,5 dB	

Chapter 9

Final Summary and Outlook

This work presents an application-specific development of integrated circuits for implantable application. A trade-off between power consumption, area, and noise was used to fulfill the acquired specification. A 130 nm process was chosen for low power consumption, high integration density and acceptable noise performance of its analogue devices. The developed ASIC was tested and evaluated in live EMG signal recording.

An implantable system for invasive EMG measurement has also been described. The implant-system acquires internal EMG signals using epimysial implantable electrodes and transmits those through a wireless link outside the body. The implant system is supplied with power using an inductive link. The system is used to investigate the benefits of an implantable approach for hand-prosthetic control. Several animal experiments (sheep, rhesus macaques) were issued to verify the functionality of the implant system and to investigate the quality of the internally recorded muscle activity compared to state-of-the-art surface recording. Muscle activity was successfully recorded from the biceps, triceps, and the deltoid muscles for over three months from a sheep and for altogether 1.5 years from two rhesus macaques. We were able to record EMG signals reliably and hence demonstrated the functionality of the system during live sessions of targeted measurements of muscle activity.

The recorded signals show high correlation with the measured force during targeted arm movement trials (Fig. 8.5). Furthermore, the internally acquired signal amplitude is almost twice as big as the externally recorded one. Despite the high noise level of internal EMG (mainly from the external energy coil and surrounding equipment) the maximum SNR and maximum information rate of the internal

EMG is higher than the external information rate acquired with commercial recording equipment. This indicates a higher efficiency of the implant system regarding data rate and hence resource consumption (area and power) compared to the external measurement system. The low-pass characteristic of the skin was noticed during the evaluation of the frequency spectrum of the internally and externally recorded signals as depicted in Fig. 8.7. The internally collected EMG is restricted to the activity of almost one muscle whereas the externally recorded EMG consists of an overlap of higher number of muscles covered by surface electrodes. This makes internal recording suitable for multiple DoF prosthetic control because of the higher selectivity and of its simplicity (no need for further processing).

9.1   Outlook

The implant system and the used ASIC succeeded in recording internal EMG activity during long targeted sessions. Despite the good quality of the recorded signals, several steps can be done to further improve the quality of the recorded signals and to optimize the implant functionality and increase its efficiency especially concerning power consumption. The improvement recommendations on chip-level (ASIC) are summarized as follows:

- **Channel gain:** The internal recorded EMG using the implant system shows huge differences of amplitude values from different muscles (high amplitude from biceps and deltoid compared to low amplitude from the triceps muscle). The placement of the epimysial electrodes and the distance to the muscular junction might explain the huge difference. The gain of the analogue channels must compensate that difference and offer the opportunity for maximum resolution per channel. Therefore, the individual gain setting is recommended for the individual channels instead of a common gain setting for all channels.

- **Multiplexer, power-down, and down-sampling:** An automatic on-chip digital control of the multiplexer and the power-down circuit reduces the number of pads and the eliminates the need for external processing through a microcontroller. Furthermore, down-sampling circuit can be implemented digitally on-chip to reduce the processing overhead through the microcontroller as described in chapter 7.1.2. The digital design must be though extensively tested to avoid the risk of malfunction.

- **Number of channels:** A higher number of recording channels is recommended to investigate the selectivity of the implanted approach by using multiple electrodes on the same target muscle. This must be combined with a strategy to reduce the overall data rate of the system.

- **Noise reduction:** The FFT of internal EMG shows a strong coupled noise from the external energy coil. This reduces SNR and the information rate of the internal recording despite the high recorded signal amplitude. Therefore, the noise content in the recorded signal must be reduced. To reduce the coupled voltage from the energy coil, the efficiency of the wireless energy transmission system must be increased and the coil-to-coil distance must be kept to a minimum. Furthermore, the internal energy coil might be placed apart from the implant electronics to reduce the coupling effect in the implant electronics and in the electrode wires. This would though increase the area of the implanted devices and increase the malfunction risk as the number of implanted modules increases. A metallic package would actually help to shield implant electronics but the noise voltage would still couple on the electrode wires. Therefore, the geometrical separation would be the best solution in order to reduce the coupled noise from the energy coil.

  A high CMRR value is also needed to suppress common noise of any source but this would cost more energy on the ASIC. Therefore, the external precautions are favored. High reduction of overall noise can be accomplished by using an active electrode. An amplifier must be placed and packaged on the epimysial electrode. However, this approach would cost more wiring (V_{dd}, ground, and signal) per electrode pair and the manufacturing and assembly process is more complicated because an ASIC must be placed and hermetically packaged at the electrode interface. Furthermore, the active electrode must work reliably under mechanical stress.

- **Power consumption** The power consumption of the post-amplifier makes up 78% of the total power consumption of the signal acquisition ASIC. A redesign is commendable by using low-power design and just caring for appropriate output driving capability for the input capacitance of the ADC. The noise can be kept at a moderate level.

The implant system worked reliably through measurement sessions for a long period of time. However, several aspects affected its functionality and were first detected after investigating the recorded EMG signals. Following aspects are recommended for future redesign of the implant system:

- **Electrode connector:** The connector between the implant electronic and the electrode wires was placed near the packaged implant. Based on its mechanical structure, the connector introduces a risk for signal quality due to continuous mechanical stress from muscle movement. Furthermore, the huge difference in packaged volume between the implant module and the connector section increases the effect of the mechanical stress as depicted in Fig. 6.17. This leads to boosted movement artifacts in the recorded signals. It is therefore recommended to get rid of the connector near the implant electronics and improve the packaged volume concerning movement artifacts by smoothing the change to suppress the effect of mechanical stress.

- **Electrode size and placement:** The placement of the electrodes under the epimysium was successfully tested during the implantation. High signal amplitudes were measured from the muscle. The placement of the electrodes on the muscle plays a significant role for the recorded signal amplitude and its frequency components. Bipolar electrodes should be placed 20 mm away from the innervation zone to minimize temporal dispersion for static contractions [115]. A precise anatomical knowledge of the underlying innervation zones is therefore necessary [116]. A surface measurement prior to implantation could further help to identify the innervation zones and to estimate the electrode position for the muscles directly under the skin.
  To increase the selectivity of the recorded EMG a small electrode size might be useful [117]. The interelectrode distance must be tested for best signal magnitude and selectivity reasons.

- **Protection circuitry:** The recording electrodes and their wires indicate a contact between an electronic system and biological tissue. To minimize the risk of a direct current from the electronic system into the body tissue, a protection circuitry can be developed between the electrode contacts. This circuit must create an open circuit when the voltage between the contacts of the electrodes exceeds a certain limit ($\approx 500~mV$) causing a damage to the body tissue according to Morimer et al. [119].

- **Impedance measurement:** The development of a circuit for live measurement of the impedance of the input electrodes by the implant electronics might be relevant in future works for the following reasons:

  1. it monitors the electrode impedance over time and the growth of body protection tissue on the electrodes.

  2. an automatic calibration of the gain factors according to the electrode impedance is possible.

  3. detection of electrode failure or breakdown during or after operation can easily be detected.

Appendix A

Appendix

A.1 Extracellular Potential

Extracellular potential indicates the potential in the fluid outside the nerve/muscle cell triggered by action potentials. There are two sites to record this potential: internally (microelectrode or implantable electrodes), externally (surface electrodes). The external recording is the most used one to measure bio-electrical activity of body organs for monitoring or healing purposes. The placement and distance to the voltage source are important in forming the recorded wave. Furthermore, there are different recording procedures used to gather biological signals. Most popular are the monopolar and the bipolar recordings. Detailed information about the extracellular potential is depicted below.

The ionic current in the extracellular medium flows in the whole surrounding conductive volume (in the whole body tissue and not on a specific path) [3]. The highest ionic current densities are across the membrane of the cell ($j_{Na} \approx -8\ Am^{-2}, j_K \approx -8\ Am^{-2}$ [2]). The ionic current flows from bio-electrical sources (membrane sites) through surrounding tissue and thus generates small potential differences (in the range of millivolt or less) between different locations within the tissue or on the body surface (Fig. A.1). Although extracellular current loops reflect the current loops of the intracellular or transmembrane current, their waveform is often completely different in magnitude, shape and timing from intracellular waveform. The following aspects deliver a brief explanation of that difference:

- Electrode displacement: positioning of the recording electrodes.

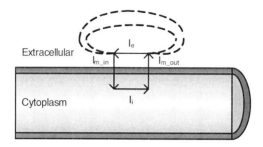

Figure A.1: Extracellular current loop. Extracellular ionic current flows along the membrane or through the surrounding conductive volume. I_e: extracellular current, I_i: intracellular current, I_{m_in} and I_{m_out}: membrane current towards the cytoplasm and towards the extracellular medium of the cell respectively.

- inter-electrode distance: the distance between the recording sites/electrodes.

- Multiple source interference: overlap of several sources of activity.

- Tissue impedance: impedance of multiple layer of surrounding tissue.

The external recording of muscle activity shows an additive overlap of muscle signals from several sources. In order to distinguish between several sources in the recorded signal the term Motor Unit (MU) is used. The motor unit consists of a motor neuron and all the muscle fibers that it innervates. A motor unit action potential (MUAP) is an additive signal of the extracellular muscle waveforms belonging to the same motor unit. The shape of the MUAP is very much dependent on the electrode displacement. The MUAP is often recorded with needle electrodes or implantable electrodes.

Appendix B

Appendix

B.1 Weak Inversion

The surface potential of a MOS transistor in weak inversion is calculated using the capacitor divider in equation B.1 assuming $V_B = 0$.

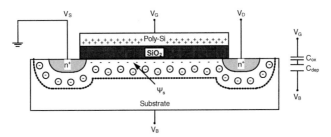

Figure B.1: MOS FET in weak inversion.

$$\psi_s = \kappa V_G \tag{B.1}$$

where κ is the gate coupling coefficient of the gate to the surface potential:

$$\kappa = \frac{C_{ox}}{C_{ox} + C_{dep}} \tag{B.2}$$

The κ is constant and has a value of approximately $\kappa \cong 0{,}6 \,..\, 0{,}8$ ($n = 1/\kappa$ is also used in some literature sources). The inversion charge in the channel correlates exponentially to the surface potential ψ_s in weak inversion. The diffusion current is proportional to the concentration gradient of particle along the channel (electron

179

concentration decreases linearly from source to drain). The drain current can then be calculated as:

$$I_D = -WD_n \frac{Q_{source} - Q_{drain}}{L} = -\frac{W}{L} \mu U_T (Q_{source} - Q_{drain}) \tag{B.3}$$

where U_T is the thermal voltage:

$$U_T = \frac{kT}{q} \cong 26.7\text{mV at body temperatur.} \tag{B.4}$$

Using the exponential relationship between the charge and the surface potential ψ we get:

$$I_D = I_0 \frac{W}{L} e^{\frac{\kappa V_G}{U_T}} \left(e^{\frac{-V_S}{U_T}} - e^{\frac{-V_D}{U_T}} \right) \tag{B.5}$$

$$I_D = I_0 \frac{W}{L} e^{\frac{\kappa V_G - V_S}{U_T}} \left[1 - e^{\frac{-V_{DS}}{U_T}} \right] \tag{B.6}$$

For $V_{DS} > 3U_T$ ($e^{-3} = 0.0498$) the last term of the equation is equal to one. The current equation is then calculated as:

$$I_D = I_0 \frac{W}{L} e^{\frac{\kappa V_G - V_S}{U_T}} \text{ for } V_{DS} > 3U_T \tag{B.7}$$

The transconductance of a MOSFET transistor in weak inversion is calculated as:

$$g_m = \frac{I_D}{V_{GS}} = \frac{\kappa I_D}{U_T} = \frac{I_D}{n U_T} \tag{B.8}$$

The factor n is between 1.2 .. 1.6 in modern technologies. The bipolar transistors possess a transconductance of $g_m = I_C/U_T$.

B.2   Operational Transconductance Amplifier

The symmetrical OTA used in the preamplifier of the MyoC1-chip is depicted in Fig 4.4. Table B.1 contains the values of the MOSFETs of the designed OTA.

B.3   MOS Bipolar Element

The MOS-bipolar element (diode-connected PMOS devices) were used in the design of the preamplifier to generate a high resistance value ($> 10^{12}$ Ω) on small silicon area. The operation of the MOS-bipolar device depend on the voltage between

Table B.1: MOSFET sizes of the preamplifier's OTA.

MOSFET device	Value W/L $[\mu m/\mu m]$
M_1, M_2	270/2.5
$M_3 - M_6$	25/25
M_7, M_8	7.2/37
M_{b1}, M_{b2}	4/4

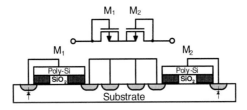

Figure B.2: MOS-bipolar element.

gate and source V_{GS}. For $V_{GS} < 0\ V$ the device acts as a diode-connected PMOS transistor. If $V_{GS} > 0\ V$ the device works as a p-n-p diode-connected BJT [120]. Two in series connected PMOS devices were sized as $W/L = 1\ \mu m/50\ \mu m$. For best operation the diode-connected devices were symmetrically connected. Two devices were used to reduce the voltage dependent resistance of the devices (Fig. B.3).

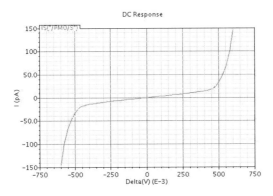

Figure B.3: Current-voltage diagram of two in series diode-connected PMOS devices.

Bibliography

[1] Jaakko Malmivuo and Robert Plonsey. *Bioelectromagnetism: Principles and applications of bioelectric and biomagnetic fields.* New York: Oxford University Press, 1995.

[2] Russell K. Hobbie and Bradley J. Roth. *Intermediate Physics for Medicine and Biology, Fourth Edition.* Springer Science+Business Media, LLC, 2007.

[3] Joseph D. Bronzino. *The Biomedical Engineering HandBook, Second Edition.* CRC Press LLC, 2000.

[4] L. Abu Saleh, W. Galjan, J. M. Tomasik, D. Schröder, and W. H. Krautschneider. Ein implantierbares system zur aufnahme von emg-signalen zur ansteuerung einer prothese. *BMT2010, Rostock, Germany*, 6(8), 2010.

[5] S. Steltenkamp, T. Dörge, R. Ruff, H. Dietl, and K.-P. Hoffmann. Electrochemical and neurophysiological testing of implantable electrode structures for the myogenic signal acquisition. *Biomed Tech*, 2010.

[6] D. Andreuccetti, R. Fossi, and C. Petrucci. An internet resource for the calculation of the dielectric properties of body tissues in the frequency range 10 hz - 100 ghz. *IFAC-CNR, Florence (Italy)*, Based on data published by C.Gabriel et al. in 1996, 1997.

[7] Statistisches Bundesamt. Statistik der schwerbehinderten menschen. *Statistisches Bundesamt, Wiesbaden 2013*, 2011.

[8] G McGimpsey and TC Bradford. Limb prosthetics services and devices critical unmet need: market analysis. *Bioengineering Institute Center for Neuroprosthetics: Worcester Polytechnic Institution*, pages 1–35, 2011.

[9] Donald G. Shurr and Thomas M. Cook. *Prosthetics & Orthotics*. Appleton & Lange A Publushing Division of Printice Hall, 1990.

[10] P Parker, K Englehart, and B Hudgins. Myoelectric signal processing for control of powered limb prostheses. *Journal of Electromyography and Kinesiology*, 16(6): 541–548, 2006.

[11] EA Clancy, Evelyn L Morin, and Roberto Merletti. Sampling, noise-reduction and amplitude estimation issues in surface electromyography. *Journal of Electromyography and Kinesiology*, 12(1):1–16, 2002.

[12] Roberto Merletti and P Di Torino. Standards for reporting emg data. *J Electromyogr Kinesiol*, 9(1):3–4, 1999.

[13] AB Simakov and JG Webster. Motion artifact from electrodes and cables. *Iranian Journal of Electrical and Computer Engineering*, 9(2), 2010.

[14] Levi J Hargrove, Kevin Englehart, and Bernard Hudgins. A comparison of surface and intramuscular myoelectric signal classification. *Biomedical Engineering, IEEE Transactions on*, 54(5):847–853, 2007.

[15] John Webster. *Medical instrumentation: application and design*. Wiley. com, 2009.

[16] H. Helmholtz. Studien ueber elektrische grenzschichten. In *Ann. Phys. Chem. 7:337-382*, 1879.

[17] David C. Grahame. The electrical double layer and the theory of electrocapillarity. *Chemical Reviews*, 41(3):441–501, 1947.

[18] F. Warburg. Ueber das verhalten sogenanter unpolarisierbarer elektroden gegen wechselstrom. *Ann. Phys. Chim.*, 67:493–499, 1899.

[19] E. Warburg. Ueber die polarisationscapacitaet des platins. *Ann. Phys.*, 6: 125–13, 1901.

[20] Leslie A Geddes and Lols E Baker. *Principles of applied biomedical instrumentation*. Wiley. com, 1968.

[21] J. E. B. Randles. Kinetics of rapid electrode reactions. part 2.-rate constants and activation energies of electrode reactions. *Trans. Faraday Soc.*, 48:828–832, 1952.

[22] Daniel P. Burbank and John G. Webster. Reducing skin potential motion artefact by skin abrasion. *Medical and Biological Engineering and Computing*, 16(1):31–38, 1978.

[23] Richard A Normann, Bradley A Greger, Paul House, Samuel F Romero, Francisco Pelayo, and Eduardo Fernandez. Toward the development of a cortically based visual neuroprosthesis. *Journal of Neural Engineering*, 6(3):035001, 2009.

[24] Il-Joo Cho, Hyoung Won Baac, and Euisik Yoon. A 16-site neural probe integrated with a waveguide for optical stimulation. In *Micro Electro Mechanical Systems (MEMS), 2010 IEEE 23rd International Conference on*, pages 995–998, 2010.

[25] T. Stieglitz, H. Beutel, R. Keller, M. Schuettler, and J.-U. Meyer. Flexible, polyimide-based neural interfaces. In *Microelectronics for Neural, Fuzzy and Bio-Inspired Systems, 1999. MicroNeuro '99. Proceedings of the Seventh International Conference on*, pages 112–119, 1999.

[26] Richard A Normann. Technology insight: future neuroprosthetic therapies for disorders of the nervous system. *Nat Clin Pract Neuro*, 3(8):444–452, August 2007.

[27] John P. Seymour and Daryl R. Kipke. Neural probe design for reduced tissue encapsulation in {CNS}. *Biomaterials*, 28(25):3594 – 3607, 2007.

[28] NEVAS M. Advances in cerebral probing using modular multifunctional probe arrays. *Medical device technology*, 18(5):38–39, 2007.

[29] W.D. Memberg, P.H. Peckham, and M.W. Keith. A surgically-implanted intramuscular electrode for an implantable neuromuscular stimulation system. *Rehabilitation Engineering, IEEE Transactions on*, 2(2):80–91, 1994.

[30] K. L. Kilgore, P. H. Peckham, M. W. Keith, K. S. Thrope, A. M. Wuolle, A. M. Bryden, and R. L. Hart. An implanted upper-extremity neuroprosthesis.

follow-up of five patients*. *The Journal of Bone & Joint Surgery*, 79(4):533–41, 1997.

[31] K.L. Kilgore, P.H. Peckham, F.W. Montague, R.L. Hart, A.M. Bryden, M.W. Keith, H. Hoyen, and N. Bhadra. An implanted upper extremity neuroprosthesis utilizing myoelectric control. In *Neural Engineering, 2005. Conference Proceedings. 2nd International IEEE EMBS Conference on*, pages 368–371, 2005.

[32] Dario Farina, Ken Yoshida, Thomas Stieglitz, and Klaus Peter Koch. Multichannel thin-film electrode for intramuscular electromyographic recordings. *Journal of Applied Physiology*, 104(3):821–827, 2008.

[33] R. Ruff, W. Poppendieck, A. Gail, S. Westendorff, M. Russold, S. Lewis, T. Meiners, and K. P Hoffmann. Acquisition of myoelectric signals to control a hand prosthesis with implantable epimysial electrodes. In *Engineering in Medicine and Biology Society (EMBC), 2010 Annual International Conference of the IEEE*, pages 5070–5073, 2010.

[34] S Lewis, H Glindemann, M Russold, S Westendorff, A Gail, T Dörge, KP Hoffmann, and H Dietl. Performance of implanted multi-site emg recording electrodes: In vivo impedance measurements and spectral analysis. *Artificial Organs*, 34(8):A46, 2010.

[35] R.F. Weir, Phil R. Troyk, G.A. DeMichele, D.A. Kerns, J.F. Schorsch, and H. Maas. Implantable myoelectric sensors (imess) for intramuscular electromyogram recording. *Biomedical Engineering, IEEE Transactions on*, 56(1):159–171, 2009. ISSN 0018-9294.

[36] J.J. Baker, E. Scheme, K. Englehart, D.T. Hutchinson, and B. Greger. Continuous detection and decoding of dexterous finger flexions with implantable myoelectric sensors. *Neural Systems and Rehabilitation Engineering, IEEE Transactions on*, 18(4):424–432, 2010. ISSN 1534-4320.

[37] Harry Nyquist. Thermal agitation of electric charge in conductors. *Physical review*, 32(1):110–113, 1928.

[38] C. Gondran, E. Siebert, S. Yacoub, and E. Novakov. Noise of surface biopotential electrodes based on nasicon ceramic and ag-agcl. *Medical and Biological Engineering and Computing*, 34(6):460–466, 1996.

[39] E. Huigen, A. Peper, and C.A. Grimbergen. Investigation into the origin of the noise of surface electrodes. *Medical and Biological Engineering and Computing*, 40(3):332–338, 2002.

[40] *NeuroNexus Technologies. (2008). Silicon Microelectrode Array Research Product Catalog & Manual.*

[41] Robert B Northrop. *Analysis and application of analog electronic circuits to biomedical instrumentation.* CRC press, 2012.

[42] H.A. Haus, W. R. Atkinson, G. M. Branch, W.B. Davenport, W. H. Fonger, W. A. Harris, S. W. Harrison, W.W. McLeod, E. K. Stodola, and T. E. Talpey. Representation of noise in linear twoports. *Proceedings of the IRE*, 48(1):69–74, 1960.

[43] Marco Pozzo. *ELECTROMYOGRAPHY (EMG), ELECTRODES AND EQUIPMENT.* Wiley, 2006.

[44] Ricardo Starbird, Wolfgang Bauhofer, Mario Meza-Cuevas, and Wolfgang H Krautschneider. Effect of experimental factors on the properties of pedot-napss galvanostatically deposited from an aqueous micellar media for invasive electrodes. In *Biomedical Engineering International Conference (BMEiCON), 2012*, pages 1–5. IEEE, 2012.

[45] Cristina Marin and Eduardo Fernández. Biocompatibility of intracortical microelectrodes: current status and future prospects. *Frontiers in neuroengineering*, 3, 2010.

[46] Louis SY Wong, Shohan Hossain, Andrew Ta, Jörgen Edvinsson, Dominic H Rivas, and H Naas. A very low-power cmos mixed-signal ic for implantable pacemaker applications. *Solid-State Circuits, IEEE Journal of*, 39(12):2446–2456, 2004.

[47] Sören Lewis, Michael Russold, Hans Dietl, Roman Ruff, Josep Marcel Cardona Audı, Klaus-Peter Hoffmann, Lait Abu Saleh, Dietmar Schroeder, Wolfgang H Krautschneider, Stephanie Westendorff, et al. Fully implantable multi-channel measurement system for acquisition of muscle activity. *IEEE TRANSACTIONS ON INSTRUMENTATION AND MEASUREMENT*, 62(7), 2013.

[48] L. Abu Saleh, W. Galjan, J. M. Tomasik, D. Schröder, and W. H. Krautschnei-
     der. Ein asic in 130nm-technologie für die aufnahme von emg-signalen zur ans-
     teuerung einer prothese. *GMM-Fachbericht-ANALOG'11*, 2011.

[49] Reid R Harrison and Cameron Charles. A low-power low-noise cmos amplifier
     for neural recording applications. *IEEE Journal of Solid-State Circuits*, 38(6):
     958–965, 2003.

[50] U Schenk and Tietze Ch. *Halbleiter-Schaltungstechnik 13. Auflage.* Springer-
     Verlag, Berlin, 2010.

[51] R. Jacob Baker. *CMOS: circuit design, layout, and simulation*, volume 18.
     Wiley-IEEE Press, 2010.

[52] Richard S. Burwen Sergio Franco Phil Perkins Marc Thompson Jim Williams
     Robert A. Pease, Bonnie Baker and Steve Winder. *Analog Circuits World Class
     Designs.* Newnes, 2008.

[53] PE Allen and DR Holberg. *CMOS Analog Circuit Design.* Oxford University
     Press, 2002.

[54] Alexander D. Mih Joni Armstrong, Judith Bell Krotoski and James W. Strick-
     land. *Hand and Upper Extremity Splinting (Third Edition) Principles & Methods.*
     Mosby, Inc. All rights reserved, 2005.

[55] Thomas A and Haddan CC. *Amputation Prosthesis.* Philadelphia: Lippincott,
     1945:1.

[56] Filer J. *Egyptian Bookshelf: Diseases.* London: British Museum, 1995:90.

[57] Putti V. *Historic Artificial Limbs.* New York: Hoeber, 1930.

[58] Little EM. *Artificial Limbs and Amputation Stumps: A Practical Handbook.*
     London: Lewis, 1922:1.

[59] Watson AB. *A Treatise on Amputations of the Extremities and their Compli-
     cations.* Philadelphia: Blakiston, 1885.

[60] Otto Bock HealthCare Gmbh. http://www.ottobock.com. *Duderstadt.*

[61] B. Hudgins, P. Parker, and R.N. Scott. A new strategy for multifunction my-oelectric control. *Biomedical Engineering, IEEE Transactions on*, 40(1):82 –94, jan. 1993.

[62] Alison Abbot. Mind-controlled robot arms show promise "people with tetraple-gia use their thoughts to control robotic aids.". *Nature*, 16 May 2012.

[63] W. Wayt Gibbs. Mind readings. *Neuroscience*, 1996.

[64] Touch Bionics Inc. http://www.touchbionics.com/.

[65] RSLSteeper BeBionic. http://bebionic.com/.

[66] W. Poppendieck and et al. Evaluation of implantable epimysial electrodes as possible interface to control myoelec-tric hand prostheses. *Technical Aids for Rehabilitation Conference*, 2011.

[67] S. Lewis, M. Hahn, C. Klein, M. F. Russold, R. Ruff, K. P. Hoffmann, E. Unger, H. Lanmueller, O. Aszmann, H. Dietl, and E. Kaniusas. Implantable silicon electrode for measurement of muscle activity : results of first in vivo evaluation. In *Biomedical Engineering / Biomedizinische Technik*, 2013, Graz, Austria.

[68] JEB Randles. Kinetics of rapid electrode reactions. *Discussions of the faraday society*, 1:11–19, 1947.

[69] Soeren Lewis. *An implantable measurement system for control of advanced arm prostheses: Electrode development, signal analysis, control algorithms and sensory feedback*. PhD thesis, Vienna University of Technology, Oct. 21 2013.

[70] Implantable medical transceiver, zl70102. URL http://www.microsemi.com/products/ultra-low-power-wireless/implantable-medical-transceivers.

[71] AJ Cardona, C Müller, R Ruff, K Becher, and KP Hoffmann. Inductive energy transmission system and real-time data link for intelligent implants. *BMT 2011, Freiburg, Germany*, pages 27.–30., September 2011.

[72] Eric R Kandel, James H Schwartz, Thomas M Jessell, et al. *Principles of neural science*, volume 4. McGraw-Hill New York, 2000.

[73] Mireya Fernandez and Ramon Pallas-Areny. Ag-agcl electrode noise in high-resolution ecg measurements. *Biomedical instrumentation & technology*, 34(2): 125–130, 2000.

[74] Y.P. Tsividis and C. McAndrew. *Operation and Modeling of the Mos Transistor*. Oxford Series in Electrical and Computer Engineering. Oxford University Press, Incorporated, 2011. ISBN 9780195170153.

[75] Andries J Scholten, Luuk F Tiemeijer, Ronald van Langevelde, Ramon J Havens, Adrie TA Zegers-van Duijnhoven, and Vincent C Venezia. Noise modeling for rf cmos circuit simulation. *Electron Devices, IEEE Transactions on*, 50 (3):618–632, 2003.

[76] Art Kay. *Operational amplifier noise: techniques and tips for analyzing and reducing noise*. Elsevier, 2012.

[77] KW Chew, KS Yeo, and S-F Chu. Impact of technology scaling on the 1/f noise of thin and thick gate oxide deep submicron nmos transistors. In *Circuits, Devices and Systems, IEE Proceedings-*, volume 151, pages 415–421. IET, 2004.

[78] Reid R Harrison. The design of integrated circuits to observe brain activity. *Proceedings of the IEEE*, 96(7):1203–1216, 2008.

[79] Paul R Gray, Paul J Hurst, Robert G Meyer, and Stephen H Lewis. *Analysis and design of analog integrated circuits*. Wiley. com, 2008.

[80] S Robinet, P Audebert, G Régis, B Zongo, JF Beche, C Condemine, et al. A low-power 0.7 vrms 32-channel mixed-signal circuit for ecog recordings. *J. On Emerging and Selected Topics in Circuits and systems*, 1(04):451–460, 2011.

[81] S. Lewis, M. Friedrich Russold, H. Dietl, R. Ruff, T. Dorge, K. Hoffmann, L. Abu Saleh, D. Schroder, W. H. Krautschneider, S. Westendorff, A. Gail, T. Meiners, and E. Kaniusas. Acquisition of muscle activity with a fully implantable multi-channel measurement system. In *Instrumentation and Measurement Technology Conference (I2MTC), 2012 IEEE International*, pages 996–999, May 2012.

[82] F. Behbahani, A. Karimi-Sanjaani, W. G. Tan, A. Roithmeier, J. C. Leete, K. Hoshino, and A. A. Abidi. Adaptive analog if signal processor for a wide-band

cmos wireless receiver. *IEEE Journal of Solid-State Circuits*, 36(8):1205–1217, 2001.

[83] Randall L Geiger and Edgar Sanchez-Sinencio. Active filter design using operational transconductance amplifiers: a tutorial. *Circuits and Devices Magazine, IEEE*, 1(2):20–32, 1985.

[84] Willy MC Sansen. *Analog design essentials*, volume 859. Springer, 2006.

[85] Arthur Bernard Williams and Fred J Taylor. *Electronic filter design handbook*, volume 15. McGraw-Hill New York, 2006.

[86] Stephen Butterworth. On the theory of filter amplifiers. *Wireless Engineer*, 7: 536–541, 1930.

[87] Randall L Geiger and John Ferrell. Voltage controlled filter design using operational transconductance amplifiers. In *Proc. IEEE/ISCAS*, pages 594–597, 1983.

[88] J. M. Tomasik, K. M. Hafkemeyer, W. Galjan, D. Schroeder, and W. H. Krautschneider. A 130nm cmos programmable operational amplifier. In *NORCHIP, 2008.*, pages 29–32. IEEE, 2008.

[89] L. Abu Saleh, W. Galjan, J. M. Tomasik, D. Schroeder, and W. H. Krautschneider. A 130nm asic for emg signal acquisition to control a hand prosthetic. In *BIODEVICES*, pages 149–153, 2012.

[90] J. Hao Cheong, K. L. Chan, P. B. Khannur, K. T. Tiew, and M. Je. A 400-nw 19.5-fj/conversion-step 8-enob 80-ks/s sar adc in 0.18-μm cmos. *Circuits and Systems II: Express Briefs, IEEE Transactions on*, 58(7):407–411, 2011.

[91] M. S. Chae, W. Liu, and M. Sivaprakasam. Design optimization for integrated neural recording systems. *Solid-State Circuits, IEEE Journal of*, 43(9):1931–1939, 2008.

[92] Elias S Greenbaum. *Implantable Neural Prostheses 2: Techniques and Engineering Approaches*, volume 2. Springer, 2010.

[93] Roberto Aparicio and Ali Hajimiri. Capacity limits and matching properties of lateral flux integrated capacitors. In *Custom Integrated Circuits, 2001, IEEE Conference on.*, pages 365–368. IEEE, 2001.

[94] Michael D. Scott, Bernhard E. Boser, and Kristofer S.-J. Pister. An ultralow-energy adc for smart dust. *IEEE Journal of Solid-State Circuits*, 38(7):1123–1129, 2003.

[95] Nathan O Sokal and Alan D Sokal. Class ea new class of high-efficiency tuned single-ended switching power amplifiers. *Solid-State Circuits, IEEE Journal of*, 10(3):168–176, 1975.

[96] Nathan O Sokal. Class-e rf power amplifiers. *QEX Commun. Quart*, (204): 9–20, 2001.

[97] S Gabriel, RW Lau, and Camelia Gabriel. The dielectric properties of biological tissues: Ii. measurements in the frequency range 10 hz to 20 ghz. *Physics in medicine and biology*, 41(11):2251, 1996.

[98] BT Bradford, W. H. Krautschneider, and D. Schroeder. Wireless power transmission for powering medical implants situated in an abdominal aortic aneurysm. *Biomed Tech*, 57:1, 2012.

[99] AJ Cardona, C Mueller, O Scholz, R Ruff, , and Germany Hoffmann, KP. Fraunhofer IBMT. Real-time data link for wireless implantable applications. 2011.

[100] Jay Han-Chieh Chang, Yang Liu, and Yu-Chong Tai. Long term glass-encapsulated packaging for implant electronics. In *Micro Electro Mechanical Systems (MEMS), 2014 IEEE 27th International Conference on*, pages 1127–1130. IEEE, 2014.

[101] W. Poppendieck, R. Ruff, A. Gail, S. Westendorff, and M. Russold. Evaluation of implantable epimysial electrodes as possible interface to control myoelectric hand prostheses. In *Technical Aids for Rehabilitation Conference*, 2011.

[102] Hans Brümmer. *Elektronische Geraetetechnik: Systematische Entwicklung und Konstruktion*. Vogel, 1980.

[103] Alexander Gail, Christian Klaes, and Stephanie Westendorff. Implementation of spatial transformation rules for goal-directed reaching via gain modulation in monkey parietal and premotor cortex. *The Journal of Neuroscience*, 29(30): 9490–9499, 2009.

[104] Martin Busch, Wolfgang Vollmann, Thomas Bertsch, Rainer Wetzler, Axel Bornstedt, Bernhard Schnackenburg, Jörg Schnorr, Dietmar Kivelitz, Matthias Taupitz, and Dietrich Grönemeyer. On the heating of inductively coupled resonators (stents) during mri examinations. *Magnetic resonance in medicine*, 54 (4):775–782, 2005.

[105] Shaobo Chen, Qingfeng Li, Weiming Wang, Bozhi Ma, Hongwei Hao, and Luming Li. In vivo experimental study of thermal problems for rechargeable neurostimulators. *Neuromodulation: Technology at the Neural Interface*, 16(5): 436–442, 2013.

[106] Peter Nordbeck, Ingo Weiss, Philipp Ehses, Oliver Ritter, Marcus Warmuth, Florian Fidler, Volker Herold, Peter M Jakob, Mark E Ladd, Harald H Quick, et al. Measuring rf-induced currents inside implants: Impact of device configuration on mri safety of cardiac pacemaker leads. *Magnetic Resonance in Medicine*, 61(3):570–578, 2009.

[107] Werner Irnich and Alan D Bernstein. Do induction cooktops interfere with cardiac pacemakers? *Europace*, 8(5):377–384, 2006.

[108] R. Buchli, P. Boesiger, and D. Meier. Heating effects of metallic implants by mri examinations. *Magnetic Resonance in Medicine*, 7(3):255–261, 1988.

[109] Taku Sato, Fumihiro Sato, and Hidetoshi Matsuki. Computer simulation of eddy current loss reduction for rechargeable cardiac pacemaker. In *Microtechnology in Medicine and Biology, 2005. 3rd IEEE/EMBS Special Topic Conference on*, pages 120–121. IEEE, 2005.

[110] J.G. Webster. Reducing motion artifacts and interference in biopotential recording. *Biomedical Engineering, IEEE Transactions on*, BME-31(12):823–826, Dec 1984.

[111] M. Milanesi, N. Martini, N. Vanello, V. Positano, M.F. Santarelli, and L. Landini. Independent component analysis applied to the removal of motion artifacts

from electrocardiographic signals. *Medical & Biological Engineering & Computing*, 46(3):251–261, 2008.

[112] Carlo J De Luca, L Donald Gilmore, Mikhail Kuznetsov, and Serge H Roy. Filtering the surface emg signal: Movement artifact and baseline noise contamination. *Journal of biomechanics*, 43(8):1573–1579, 2010.

[113] D. McDonnall, S. Hiatt, C. Smith, and K. S. Guillory. Implantable multichannel wireless electromyography for prosthesis control. In *Engineering in Medicine and Biology Society (EMBC), 2012 Annual International Conference of the IEEE*, pages 1350–1353. IEEE, 2012.

[114] S Lewis, M Hahn, C Klein, MF Russold, R Ruff, KP Hoffmann, E Unger, H Lanmüller, O Aszmann, H Dietl, et al. Implantable silicone electrode for measurement of muscle activity: Results of first in vivo evaluation. *Biomedical Engineering/Biomedizinische Technik*, 2013.

[115] G. Kamen and D. A. Gabriel. *Essentials of electromyography*. Champaign, IL: Human Kinetics, 2010.

[116] A. Perotto and E.F. Delagi. *Anatomical Guide for the Electromyographer: The Limbs and Trunk*. Charles C Thomas, 2005. ISBN 9780398075774.

[117] Isak Gath and Erik V Stalberg. Techniques for improving the selectivity of electromyographic recordings. *Biomedical Engineering, IEEE Transactions on*, (6):467–472, 1976.

[118] I Gath and Erik Stalberg. Measurements of the uptake area of small-size electromyographic electrodes. *Biomedical Engineering, IEEE Transactions on*, (6):374–376, 1979.

[119] J Thomas Mortimer, David Kaufman, and Uros Roessmann. Intramuscular electrical stimulation: tissue damage. *Annals of biomedical engineering*, 8(3): 235–244, 1980.

[120] Tobi Delbruck and Carver A Mead. Adaptive photoreceptor with wide dynamic range. In *Circuits and Systems, 1994. ISCAS'94., 1994 IEEE International Symposium on*, volume 4, pages 339–342. IEEE, 1994.

List of Publications

[own1] **L. Abu Saleh**, Jakob M. Tomasik, Wjatscheslaw Galjan, and Wolfgang H. Krautschneider. A SoC Based Mobile EEG Signal Acquisition System Using Multi-Sensor-Recording to Reduce Noise and Artifacts. In *ProR-ISC2008 Workshop*, 11 2008.

[own2] **L. Abu Saleh**, W. Galjan, J. M. Tomasik, D. Schröder, and W. H. Krautschneider. Ein implantierbares system zur aufnahme von emg-signalen zur ansteuerung einer prothese. *BMT2010, Rostock, Germany*, 6 (8), 2010.

[own3] Arthur Schott, **L. Abu Saleh**, Jakob Tomasik, Wjatscheslaw Galjan, , and Wolfgang Krautschneider. Design measures for performance compensation of analog circuits during reduction of technology structures. In *ProRISC2010 Program for Research on Integrated Circuits and Systems*, 11 2010.

[own4] S. Lewis, M. Russold, H. Dietl, S. Westendorff, A. Gail, T. Dörge, R. Ruff, K.-P. Hoffmann, **L. Abu Saleh**, D. Schröder, and W. Krautschneider. Detection of arm movement from EMG signals recorded with fully implanted electrodes: A case study in a rhesus macaque. In *BMT 2011*, 09 2011.

[own5] Mario Meza, **L. Abu Saleh**, Dietmar Schroeder, and Wolfgang Krautschneider. ASIC for Neurostimulation with Different Waveforms. In *BMT 2011*, 09 2011.

[own6] **L. Abu Saleh**, W. Galjan, J. M. Tomasik, D. Schröder, and W. H. Krautschneider. Ein asic in 130nm-technologie für die aufnahme von emgsignalen zur ansteuerung einer prothese. *GMM-Fachbericht-ANALOG'11*, 2011.

[own7] Mario A Meza Cuevas, **L. Abu Saleh**, D. Schroeder, and W. H. Krautschneider. Toward the optimal architecture of an asic for neurostimulation. In *BIODEVICES*, pages 179–184, 2012.

[own8] S. Lewis, M. Friedrich Russold, H. Dietl, R. Ruff, T. Dorge, K. Hoffmann, **L. Abu Saleh**, D. Schroder, W. Krautschneider, S. Westendorff, A Gail, T. Meiners, and E. Kaniusas. Acquisition of muscle activity with a fully implantable multi-channel measurement system. In *Instrumentation and Measurement Technology Conference (I2MTC), 2012 IEEE International*, pages 996–999, May 2012.

[own9] **L. Abu Saleh**, W. Galjan, J. M. Tomasik, D. Schroeder, and W. H. Krautschneider. A 130nm asic for emg signal acquisition to control a hand prosthetic. In *BIODEVICES*, pages 149–153, 2012.

[own10] S. Lewis, M. Russold, H. Dietl, R. Ruff, J.M.C. Audi, K.-P. Hoffmann, **L. Abu Saleh**, D. Schroeder, W.H. Krautschneider, S. Westendorff, A Gail, T. Meiners, and E. Kaniusas. Fully implantable multi-channel measurement system for acquisition of muscle activity. *Instrumentation and Measurement, IEEE Transactions on*, 62(7):1972–1981, July 2013.

[own11] K.-P. Hoffmann, **L. Abu Saleh**, J.M. Cardona Audi, H. Dietl, H. Frank, A. Gail, E. Kaniusas, W.H. Krautschneider, S. Lewis, T. Meiners, R. Ruff, M. Russold, D. Schroeder, and S. Westendorff. Implantierbares myoelektrisches Assistenzsystem zur intuitiven Steuerung einer bionischen Handprothese. In *OTWorld 2014*, 05 2014.

Lebenslauf

Name	Abu Saleh
Vorname	Lait
Geburtsdatum	19.06.1980
Geburtsort, -land	Majdal-Shams (Golanhöhen), Syrien

Schule und Studium

09.1994 - 06.1998	Gymnasium in Majdal-Shams (Golanhöhen)
10.2001 - 05.2008	Studium an der Technischen Universität Hamburg-Harburg, Studienrichtung: Informatikingenieurwesen
05.2008	Diplom-Ingenieur

Berufstätigkeit

06.2004 - 07.2005	Studentischer Mitarbeiter an der Technischen Universität Hamburg-Harburg:
	- School-Lab (Arbeitsbereich Mathematik, TUHH)
	- Tutor für Digital Entwurfspraktikum (VHDL)
06.2006 - 04.2008	Hardwarenahe Programmierung von Überwachungssystemen, BMC IT Health Solutions, Hamburg

08.2008 - 08.2011	Wissenschaftlicher Mitarbeiter im Institut für Nano- und Medizinelektronik für das BMBF-Projekt **"Myo-Plant"** (16SV3699)
09.2011 - 02.2012	Wissenschaftlicher Mitarbeiter an der Technischen Universität Hamburg-Harburg im Rahmen des Projektes Lehrinnovation: Entwicklung, Einführung und Umsetzung von PBL in der Lehrveranstaltung **"Medizinelektronik"**
03.2012 - 02.2014	Wissenschaftlicher Mitarbeiter im Institut für Nano- und Medizinelektronik für das BMBF-Projekt **"Myo-Plant"** (16SV3699)
03.2014 - 06.2014	Wissenschaftlicher Mitarbeiter im Institut für Nano- und Medizinelektronik für das BMWi-Projekt **"ESi-MED"**
Seit 07.2014	Wissenschaftlicher Mitarbeiter im Institut für Nano- und Medizinelektronik für das DFG-Projekt "Hochauflösende Charakterisierung der funktionellen Konnektivität und des Verhaltens in gesunden und transgenen Tieren von der Neugeborenenperiode bis zum Erwachsenenalter"

Bisher erschienene Bände der Reihe

Wissenschaftliche Beiträge zur Medizinelektronik

ISSN 2190-3905

Alle erschienenen Bücher können unter der angegebenen ISBN-Nummer direkt online
(http://www.logos-verlag.de) oder per Fax (030 - 42 85 10 92) beim Logos Verlag
Berlin bestellt werden.